Advanced Mathematical Techniques in Science and Engineering

RIVER PUBLISHERS SERIES IN MANAGEMENT SCIENCES AND ENGINEERING

Indexing: All books published in this series are submitted to the Web of Science Book Citation Index (BkCI), to CrossRef and to Google Scholar.

The "River Publishers Series in Management Sciences and Engineering" looks to publish high quality books on management sciences and engineering. Providing discussion and the exchange of information on principles, strategies, models, techniques, methodologies and applications of management sciences and engineering in the field of industry, commerce and services, it aims to communicate the latest developments and thinking on the management subject world-wide. It seeks to link management sciences and engineering disciplines to promote sustainable development, highlighting cultural and geographic diversity in studies of human resource management and engineering and uses that have a special impact on organizational communications, change processes and work practices, reflecting the diversity of societal and infrastructural conditions.

The main aim of this book series is to provide channel of communication to disseminate knowledge between academics/researchers and managers. This series can serve as a useful reference for academics, researchers, managers, engineers, and other professionals in related matters with management sciences and engineering.

Books published in the series include research monographs, edited volumes, handbooks and text books. The books provide professionals, researchers, educators, and advanced students in the field with an invaluable insight into the latest research and developments.

Topics covered in the series include, but are by no means restricted to the following:

- Human Resources Management
- Culture and Organisational Behaviour
- Higher Education for Sustainability
- SME Management
- Strategic Management
- Entrepreneurship and Business Strategy
- Interdisciplinary Management
- Management and Engineering Education
- Knowledge Management
- Operations Strategy and Planning
- Sustainable Management and Engineering
- Production and Industrial Engineering
- Materials and Manufacturing Processes
- Manufacturing Engineering
- Interdisciplinary Engineering

For a list of other books in this series, visit www.riverpublishers.com

Advanced Mathematical Techniques in Science and Engineering

Editors

Mangey Ram

Graphic Era (Deemed to be University)
India

João Paulo Davim

University of Aveiro
Portugal

Published 2018 by River Publishers
River Publishers
Alsbjergvej 10, 9260 Gistrup, Denmark
www.riverpublishers.com

Distributed exclusively by Routledge
4 Park Square, Milton Park, Abingdon, Oxon OX14 4RN
605 Third Avenue, New York, NY 10017, USA

Advanced Mathematical Techniques in Science and Engineering / by Mangey Ram, João Paulo Davim.

Routledge is an imprint of the Taylor & Francis Group, an informa business

ISBN 978-87-93609-34-1 (print)

Contents

Preface

In recent years, mathematical techniques have an extraordinary growth in science and engineering to all novel disciplines. The book Advanced Mathematical Techniques in Science and Engineering engrossed on a comprehensive range of mathematics applied in various fields of science and engineering for different tasks.

Chapter 1 presents the results of statistical modeling describing the dependence of the time to reach consensus on the number and authoritarianism of a social group members using two mathematical models of consensus achievement in a group based on the model proposed by DeGroot and model of the cellular automaton.

Chapter 2 provides the problems of classification and quantitative assessment of intersystem accident consequences including cascade failure process. The research conducted has made it possible to classify the intersystem failures occurring in critical infrastructures. The classification of ISFs by the structure of development of abnormal processes is of the greatest interest.

Chapter 3 gives the description of capabilities of applied mathematics in social and military crisis management. Crisis management of environment, infrastructure, migration, and some other issues of homeland security requires a correct base for decision-making.

Chapter 4 analyses a scenario involving processing at Access Point (AP) is modeled in network simulator (NS2). The computational flooding DoS attack is successful via data frame flooding on AP and the attack is able to disrupt the wireless station communication. Key Hiding Communication (KHC) scheme is proposed as a secure and lightweight communication for WLANs.

Chapter 5 presents the two methods of ranking objects used to make a decision if the situation is unclear. The first part of this chapter examines the method of putting the objects arranged within homogeneous groups in accordance with a specified criterion into a general ranked list and the second part observes the method that helps reconstruct the preferences experts had as far as the method of analyzing hierarchies is concerned if for some reason experts could not compare the pairs of objects.

Chapter 6 investigates the influence of different nature of information flow on the diffusion of a product. Four different models under varied nature of awareness and adoption are modeled and empirically tested on two different consumer durables. The obtained results show good consistency on both datasets.

Chapter 7 discusses various soft computing techniques used in science and engineering and presents a survey of a range of applications of these fields ranging from the purely theoretical to the most practical ones.

Chapter 8 provides the various aspects related to reliability analysis and reliability optimization.

Chapter 9 discusses the application of tree growth models for forest ecosystems modelling and results that there is need for close interaction between modelers and forestry practitioners for mutual appreciation of methods and approaches used for ecological modeling.

This research book can be used as a support book for the final undergraduate engineering course (for example, mechanical, mechatronics, industrial, computer science, information technology etc.). Also, this book can serve as a valuable reference for academics, mechanical, mechatronics, computer science, information technology and industrial engineers, environmental sciences as well as, researchers in related subjects.

Mangey Ram

J. Paulo Davim

Acknowledgements

The Editors acknowledges River Publishers for this opportunity and professional support. Also, we would like to thank all the chapter authors and reviewers for their availability for this work.

List of Contributors

Aronov Iosif Zinovievich, *Research Center, International Trade and Integration, Moscow, Russia*

Adarsh Anand, *Department of Operational Research, University of Delhi, India*

Alexander V. Bochkov, *Risk Analysis Center, NIIgazeconomika, LLC, Moscow, Russia*

Anuj Kumar, *Department of Mathematics, University of Petroleum and Energy Studies, India*

Deepti Aggrawal, *Department of Operational Research, University of Delhi, India*

Grigoryev Vadim Iosifovich, *Moscow State University, Moscow, Russia*

Ivan Kopachevsky, *Centre for Aerospace Research of the Earth of National Academy of Sciences of Ukraine, Ukraine*

Lata Nautiyal, *Department of Computer Applications, Graphic Era (Deemed to be University), India*

Maksimova Olga Vladimirovna, *National Research University, Higher School of Economics, Moscow, Russia*

Mangey Ram, *Department of Mathematics, Computer Science and Engineering, Graphic Era (Deemed to be University), India*

Maxim Yuschenko, *Centre for Aerospace Research of the Earth of National Academy of Sciences of Ukraine, Ukraine*

Mohini Agarwal, *School of Business, Galgotias University, India*

Nikolay N. Zhigirev, *Risk Analysis Center, NIIgazeconomika, LLC, Moscow, Russia*

Preeti Malik, *Department of Computer Science and Engineering, Graphic Era (Deemed to be University), India*

Rajeev Singh, *Department of Computer Engineering, G. B. Pant University of Agriculture and Technology, Pantnagar, India*

Rajesh Joshi, *G. B. Pant National Institute of Himalayan Environment and Sustainable Development, Kosi-Katarmal, Almora, India*

Rubina Mittal, *Keshav Mahavidyalaya, University of Delhi, India*

Rupesh Dhyani, *G. B. Pant National Institute of Himalayan Environment and Sustainable Development, Kosi-Katarmal, Almora, India*

Sangeeta Pant, *Department of Mathematics, University of Petroleum and Energy Studies, India*

Tatiana B. Timofeeva, *The State University of Management, Moscow, Russia*

Teek Parval Sharma, *Department of Computer Engineering, National Institute of Technology, Hamirpur (H.P.), India*

Valery V. Lesnykh, *The National Research University Higher School of Economics, Energy Institute, Moscow, Russia*

Vladislav S. Petrov, *NIIgazeconomika, LLC, Moscow, Russia*

Yuriy V. Kostyuchenko, *Centre for Aerospace Research of the Earth of National Academy of Sciences of Ukraine, Ukraine*

List of Figures

List of Tables

List of Abbreviations

ACK	Acknowledgement
AP	Access Point
APEC	Asia-Pacific Economical Cooperation
BGC-Models	Biogeochemical models
CBR	Constant bit Rate
CD	Codeword
CTS	Clear To Send
DIFS	Distributed Inter-Frame Space
DoS	Denial of Service
DST	Department of Science and Technology
FTP	File Transfer Protocol
GBPNIHESD	G. B. Pant National Institute of Himalayan Environment and Sustainable Development
GDP	Gross Domestic Product
GIS	Geographic Information System
IEC	International Electrotechnical Commission
IEEE	Institute of Electrical and Electronic Engineers
ISF	Intersystem failure
ISO	International Organization for Standardization
IV	Initial Vector
KHC	Key Hiding Communication
KPCA	Kernel-based Principal Component Algorithm
L\DPR	Lugansk and Donetsk People's Republics
LAN	Local Area Network
MAC	Medium Access Control
MI	Mutual Information
MIC	Message Integrity Code
MOD	Modulus Operator
MRF	Markov Random Field
NASA	National Aeronautics and Space Administration
NDVI	Normalized Difference Vegetation Index

NDWI	Normalized Difference Water Index
NIR	Near Infra-Red
NMI	Normalized Mutual Information
NMSHE	National Mission for Sustaining the Himalayan Ecosystem
NPP	Net Primary Productivity
NS	Network Simulator
RCS	Regional Curve Standardization
RTS	Request To Send
SIFS	Short Inter-Frame Space
STA	Wireless Station
SWIR	Short-Wave Infra-Red
TCs	Technical committees on standardization
TCP	Transmission Control Protocol
TF-3	Task Force 3
UDP	User Datagram Protocol
VIIRS	Visible Infrared Imaging Radiometer Suite
WEP	Wired Equivalent Privacy
Wi-Fi	Wireless Fidelity
WLAN	Wireless LAN
WPA	Wi-Fi Protected Access
XOR	Exclusive OR

1

Analysis of Consensus-Building Time in Social Groups Based on the Results of Statistical Modeling

Aronov Iosif Zinovievich[1], Maksimova Olga Vladimirovna[2] and Grigoryev Vadim Iosifovich[3]

[1]Research Center, International Trade and Integration, Moscow, Russia
[2]National Research University, Higher School of Economics, Moscow, Russia
[3]Moscow State University, Moscow, Russia

The problem of studying the achievement of consensus in social groups is related to the complexity of organizing such a study, especially for large groups, with more than five participants. In connection with the above, it is advisable to investigate the phenomenon of consensus in large social groups, using the modeling methodology.

The article presents the results of statistical modeling describing the dependence of the time to reach consensus on the number and authoritarianism of a social group members using two mathematical models of consensus achievement in a group based on the model proposed by DeGroot and model of the cellular automaton.

The main problems of attaining consensus under the settings of the proposed model during the development of consensus standards in technical committees on standardization were analyzed. It is shown that an increase in the number of social group members and their authoritarianism has an adverse impact on the time to reach consensus and increases the disunity of the group.

A model of the cellular automaton modeling the achievement of consensus within the negotiation process has been studied: the initial discrepancy between the opinions of the members of the group and the space of opinions of the members of the group. In particular, it is shown that if initially the views of the members of the group are radically different, then the process of reaching consensus will be as long as possible if one of the participants is absolutely authoritative. If initially the views of the members of the group are close, then the process of reaching consensus will also be as long as possible if both members of the group are absolutely compromise.

1.1 Introduction and Purpose of the Study

Consensus as a method of decision-making is often used in the activities of social groups. In the context of this article, by social group we mean a set of people of a fixed number that interact with each other to achieve a specific goal. In this regard, the passengers of an aircraft are not a social group, and negotiators can be considered as a social group. Typical examples of such social groups are technical committees on standardization (TCs) developing voluntary consensus standards, for example, ISO (International Organization for Standardization) or IEC (International Electrotechnical Commission) standards.

In this study we consider consensus as the absence of fundamental disagreement among the members of a social group regarding the issue under consideration.

The principle of consensus-based decision-making (in short consensus) is currently widely used in various dialogues (bilateral or multilateral negotiations), in the work of international organizations (e.g. APEC (Asia-Pacific Economical Cooperation), ISO, IEC), in the activities of parties and political organizations, when developing medical recommendations.

Comparing the consensus as a method of decision-making with voting, it should be noted that voting initially generates competition (and not cooperation), does not take into account the possibility of compromise, forces the minority to submit to the majority opinion (which in fact does not happen, because the minority as a rule remains unconvinced), and disturbs cohesion of a society or a group [1]. The problems of reaching a consensus in a social group, which is based, as a rule, on ability and capacity of its members to find a compromise, have been poorly investigated.

The main problem of studying the achievement of consensus in social groups is related to the complexity of organizing such a study, especially for large groups, with more than five participants. Therefore, it is advisable to investigate the phenomenon of consensus in large social groups using the modeling methodology.

The article of [2] demonstrated the principal possibility of describing the process of achieving consensus based on regular Markov chains [3]. Recently, this model has been applied in different fields, for example, in network automation management [4], in negotiation process [5].

In addition, in the interest of the study, it is advisable to use a different approach, based on the methodology of cellular automata [6].

In practice, it is important to assess the speed of the convergence of experts' opinions in relation to the parameters of various social groups: the number of members, their authoritarianism[1] and the magnitude of discrepancies of opinions of members group, which has not been analyzed so far.

The purpose of the article is to investigate the impact of these factors on the time to reach a consensus using the apparatus of regular Markov chains that satisfactorily describes the process of a consensus-building and methodology of cellular automata.

1.2 Description of the Model for Consensus Based on Regular Markov Chains

Let us denote by n the number of members of the group participating in the discussion. Let S(0) = ($s_0{}^1$; . . . ; $s_0{}^n$) be the vector of initial opinions of the members of the group, where $s_0{}^i$ is the initial opinion of the i-th expert. Members of the group (experts) exchange views on the values of the vector S during the group meeting. At the same time, the opinion of each expert may vary depending on the degree (level) of the trust of this expert to the opinion of the other member of the group, and also on the degree of self-confidence.

Along with the activity of the discussion participants, consensus often requires flexibility in decision-making. In this regard, we set the probability of the trust of the i-th expert in the opinion of the j-th expert by the strict inequality $0 < p_{ij} < 1$ ($i = 1, \ldots, n; j = 1, \ldots, n, i \neq j$). It is also assumed that the i-th expert trusts himself with a certain probability of $0 < p_{ii} < 1$. They can also be interpreted as levels of authoritarianism among experts: the higher the value of p_{ii}, the higher the authoritarianism of the member of the group. Questions of authoritarian personality have been studied in sufficient detail, which allows us to introduce a corresponding scale [7]. This is important for practical recommendations on group management.

---

[1] **Authoritarianism** [from the Latin *autoritas* – influence, power] – social and psychological feature of an individual that reflects the desire to maximally submit the interaction and communication partners.

Thus, to model the group activities, we will take into account the loyalty of experts, abandoning absolutely authoritarian members and absolutely irresponsible ones in making decisions. As a result of modeling, a square matrix of trust $\mathbf{P}_{nxn} = (p_{ij})$ is formed which defines a sequential process of coordinating the views of the group members. The sum of the probabilities p_{ij} in each row of the matrix is 1, that is,

$$\forall i \in \overline{1,\ n} \left[\sum\nolimits_{j=1}^{n} p_{ij} = 1 \right].$$

At the first step in coordinating the opinions of experts, the vector of opinions of group members is calculated by the formula

$$\mathbf{S}^{\mathrm{T}}(1) = \mathbf{P}_{n \times n} \cdot \mathbf{S}^{\mathrm{T}}(0) = (\mathrm{s}_1^1, \ldots, \mathrm{s}_1{}^n)^{\mathrm{T}},$$

where $\mathbf{S}^{\mathrm{T}}(\cdot)$ is a column vector of dimension $n \times 1$.

After the *k*-th step of approvals, the vector of opinions is calculated by the formula.

$$\mathbf{S}^{\mathrm{T}}(k) = (\mathrm{s}_k{}^1, \ldots, \mathrm{s}_k^n)^{\mathrm{T}} = \mathbf{P}_{n \times n} \cdot \mathbf{S}^{\mathrm{T}}(k-1) = \mathbf{P}_{n \times n}^k \cdot \mathbf{S}^{\mathrm{T}}(0). \qquad (1.1)$$

The iterative process stops at the *m*-th step if all rows of the matrix $\mathrm{P}_{nxn}{}^m$ become the same.

Mathematically, this means that the confidence matrix $\mathbf{P}_{nxn}$ after *m* iterations has reached the final matrix **F**, and since the final matrix **F** does not change with subsequent iterations, the expert opinion vector S(*m*) = S(0), $\mathbf{P}_{nxn}{}^m = (\mathrm{s}_m{}^1, \ldots, \mathrm{s}_m{}^n)$ does not change accordingly. This agrees with the well-known concept of the group dynamics theory describing the processes that take place in social groups [8].

The theory of Markov chains [3] suggests that the necessary and sufficient condition for the convergence of the initial matrix **P** to the final matrix **F** (the necessary and sufficient condition for reaching consensus) for any vector of initial opinions is the regularity[2] of the matrix **P**. In other words, it is necessary and sufficient that the sum over the rows of the matrix **P** is equal to 1 and that for any probabilities p_{ij} the strict inequality $0 < p_{ij} < 1$ is fulfilled. As it was written above, in terms of social group activity it is important that the members are loyal.

[2]The matrices, for which the sums of elements of all rows are equal to 1, are called stochastic. If for some *n* all elements of the matrix $\mathbf{P}^n$ are not equal to 0, then such a transition matrix is called regular.

Thus, if the matrix of trust **P** is regular (that is, there are not ambitious experts with a distinct opinion), then whatever the initial views of the group members were, consensus is achievable, although it may require a significant number of iterations (discussions in the social group). This means that in some cases, even with loyal experts, considerable time is required to achieve it.

1.3 Specific Cases in the Model of Attaining Consensus in the Work of TC

The fact that the considered model is workable is demonstrated by the analysis of situations in which the condition of loyalty of experts is violated [9].

1.3.1 Domination

If there is one authoritative expert ($\exists i = \overline{1, n\ }; p_{ii} = 1$) in the group, then his opinion as a result of the approvals (iterations) does not change (in the final matrix **F**, the element p_{ii} remains equal to 1). The presence in the TC of both an authoritative expert and an expert with a level of self confidence close to 1 delays consensus for a long time. Indeed, such a member of the TC is difficult to convince. Therefore, the inclusion of an ambitious member in TC should be strongly checked, since the opinion of this particular expert will prevail. For this reason, for example, representatives of authorities in the TC should be only as ordinary members of committees.

1.3.2 Presence of Several Leaders

The situation when there are several leaders in a group is characterized by a matrix **P** in which several entries on the main diagonal are equal to 1. Matrices of this type and the Markov chains corresponding to them are said to be decomposable [3][3]. In this situation, consensus is never achievable (for any $n > 2$). In the literature on group dynamics, similar conclusions are made [10].

The presence of several leaders in a group fundamentally distinguishes this situation from the previous one. The presence of one leader in a group provides a consensus which would be of low quality however in terms of

[3]A matrix A is said to be decomposable if by rearrangements of rows it can be reduced to the form $\tilde{A} = \begin{pmatrix} B & 0 \\ C & D \end{pmatrix}$, where B and D are square matrices.

the required numbers of approvals, whereas the presence of several leaders in a group leads to a total impossibility of consensus. This was confirmed by a sufficiently large number of observations obtained during monitoring of various groups work: the more ambitious members in the TC, the more difficult it is to attain consensus in the group [10].

1.3.3 Global Domination

If in the group all experts have a high self-esteem (i.e. we can assume that $\forall i = \overline{1;n} p_{ii} = 1$), then the trust matrix **P** in this case becomes a matrix of 1's. Since for any number of *m* iterations (TC discussions) $\mathbf{P}^m = \mathbf{E}^m = \mathbf{E}$, the matrix **P** does not converge to the final one, and consequently the consensus in this case in principle is unattainable.

1.3.4 Responsibility Shift

The situation in which each expert shifts full responsibility to another member of the group, escaping responsibility for making a decision: experts join the opinion of the group, considering it to be correct and their assessment to be erroneous. In the theory of Markov chains it is shown that the corresponding transition matrix does not converge to the final matrix [3]. Consequently, for such a group, consensus is not achievable. In fact, at least two "irresponsible" experts will be enough not to reach consensus in a group. The analysis of group dynamics shows that this situation is very common in life.

1.3.5 Coalitions

This is another case of the impossibility of a consensus, which is associated with the formation of coalitions within the group. The corresponding matrices are decomposable, in this situation a consensus cannot be reached for any $n > 2$. In the literature on group dynamics, similar situations are considered and conclusions are given [10]. The task of the group leader is to eliminate the existing group coalitions through the choice of compromise solutions.

1.4 Analysis of the General Case in the Consensus Model

Let us evaluate the impact of the number of group members and their authoritarianism on the time of consensus-building using statistical modeling. The influence of the number *n* of group members on the time to reach consensus was investigated using statistical modeling. As a dependent variable, *m* is the number of discussions in the group before consensus is reached, at which the condition is met.

$$\det \mathbf{P}^m < \varepsilon. \tag{1.2}$$

The modeling involved several stages:

The first stage is the choice of factor 1 levels (the number of group members, n) for practical reasons: 5; 10; 20; 30; 40; 50. It seems that the chosen boundaries of the modeling parameters correctly describe the actual situation in groups.

The second stage is the selection of factor 2 levels (probabilities p_{ii}, which show the expert's self-confidence): 0.20–0.30; 0.45–0.55; 0.65–0.75; 0.85–0.95; 0.90–1.00. In reality, the group members combine the behavioral implications of all groups (from authoritarian to conformist), which is reflected in the simulation conditions $0.2 < p_{ii} < 1$.

The third stage – for each level n, the diagonal elements p_{ii} of the matrix **P** were modeled according to the uniform distribution law; the parameters of the uniform law were the boundaries of the corresponding level of the factor p_{ii}. In each row of the matrix **P**, the residual probabilities p_{ij} ($i \neq j$) were also modeled according to a uniform law with parameters 0 and 1 and normalized so that the sum of the probabilities within each row was equal to 1, that is, the matrix **P** became stochastic.

The accuracy of the matrix elements modeling was 0.01, which was derived from the considered boundaries of the number n of members in the group. The values of ε in the inequality (Equation 1.2) were determined by the accuracy of modeling of the elements in the matrix **P** and its size n:

$$\varepsilon = \left(\frac{0,01}{n-1}\right)^n, \quad n \geq 2.$$

This type of dependence of ε on the number of members n was derived from:

- the change of accuracy of (n – 1) element in each row of the matrix under normalization;
- the power-law dependence related to the technique of calculating the determinant of the n-th order.

To obtain consistent conclusions on average number of approvals m while changing other parameters, 100 simulations were carried out in Excel at each fixed level of factors n and p_{ii} [11]. We analyzed both the impact of authoritarianism of the group members and their number on the number of meetings necessary to reach a consensus. A three-dimensional model of the dependence is shown in Figure 1.1, and two-dimensional dependencies are presented in Figures 1.2 and 1.3.

It was found that the most obvious was the relationship between the number of approvals and the level of authoritarianism p_{ii} of the group members for a fixed number of members (Figure 1.3). The graphs clearly illustrate the high sensitivity of the number of approvals m to the growth of authoritarianism, especially stating from p_{ii} equal to approximately 0.8.

With an increase in the number of committee members, a gradual power-law rise of the number of approvals was observed (Figure 1.2). The analysis of the model confirmed not only the visual, but also theoretically good agreement with the data (for each curve $R^2 \approx 0.97$). The study shows that the increase in the number of members of the group (with a fixed level of authoritarianism of members) leads to the increase in the number of iterations m to reach consensus (Figure 1.2). At the same time, in the case of a low level of authoritarianism ($p_{ii} < 0.5$) its influence decreases.

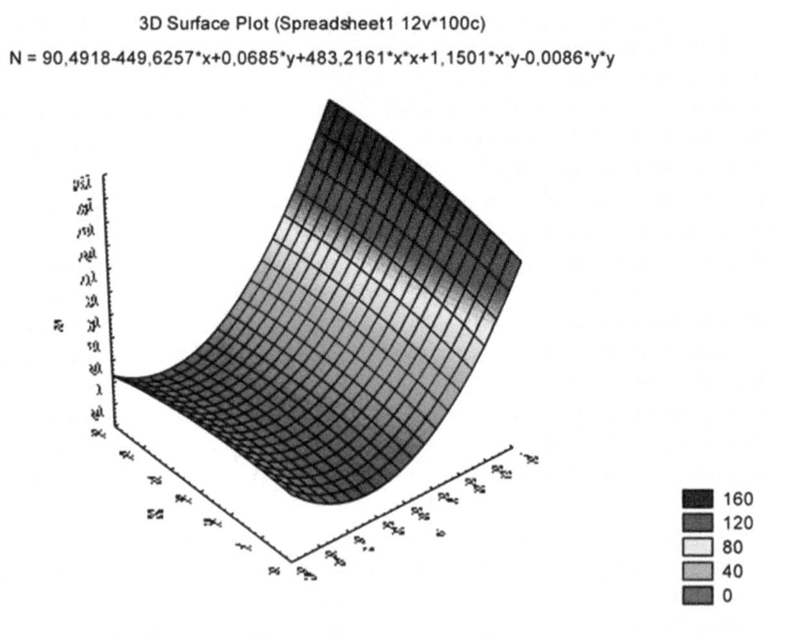

Figure 1.1 Regression model of the dependence of the number of approvals before reaching consensus on the number of members (the left axis) and the level of authoritarianism of members (the right axis).

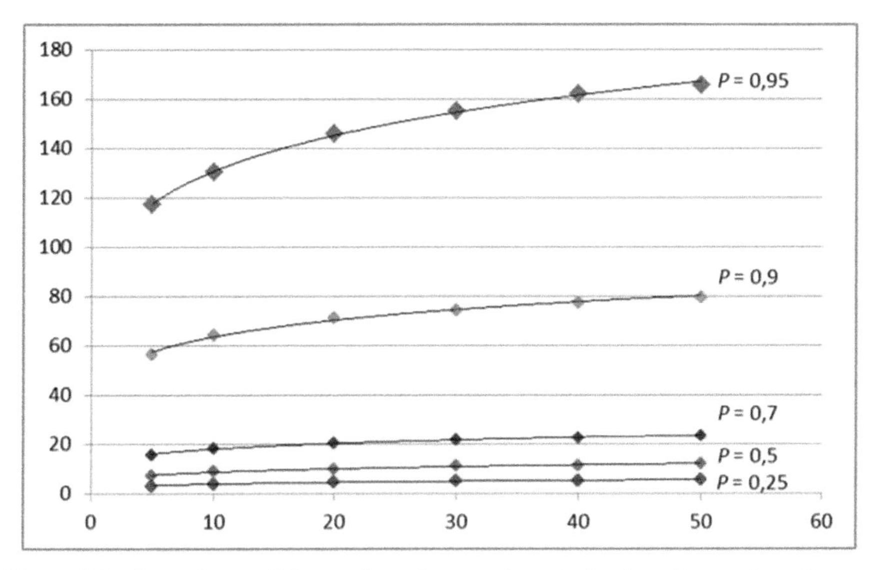

Figure 1.2 Dependence of the number of approvals m (axis y) on the number of group members n (axis x): the corresponding average values of the authoritarian level $\mathbf{P} = p_{ii}$ are shown next to the curves.

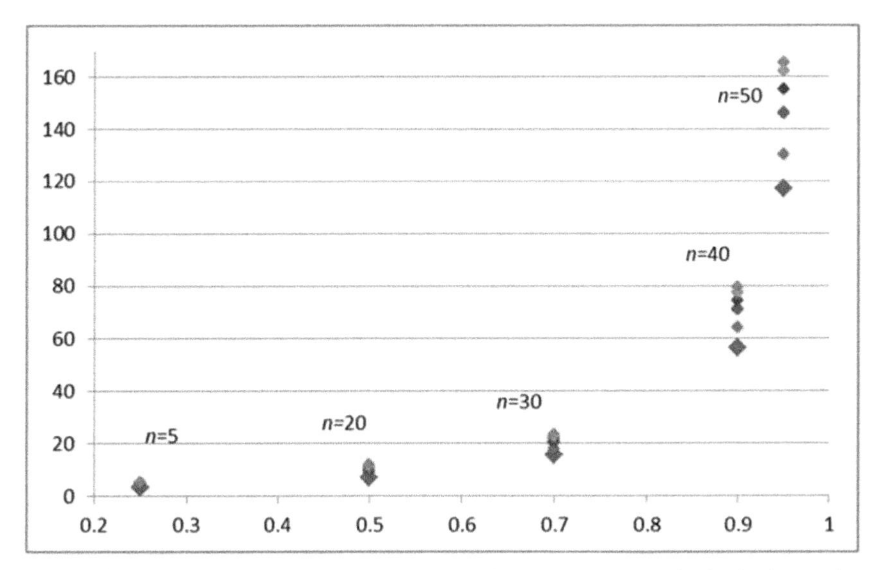

Figure 1.3 Dependence of the number of approvals m (axis y) on authoritarianism (axis x) under a fixed number of group members n.

Another specific fact attracts the attention: with the increase in the number of members of the group and the increase in the authoritarianism of its members, the spread of the numbers of approvals m to attain consensus (relative to average values) on average increases, which indicates growth of the disunity in the expert group.

1.5 Management of the TCs by the National Standardization Body

Since an increase in the number of members of a TC and/or an increase in the authoritarianism of TC members leads to an increase in the number of steps (iterations) m necessary for the convergence of the trust matrix $\mathbf{P}_{(n \times n)}$ to the final matrix **F** (in other words, to the growth of the time of a consensus building) obviously, this situation cannot satisfy a national standardization body (center). Therefore, it is necessary to introduce control (management) from the center. The target function of the center in relation to this situation can be written as follows:

$$\mathrm{Z}(m) = m < \mathrm{M}, \tag{1.3}$$

where M is the planned number of iterations. If we match each approval by the TC with a unit of time, then the inequality (Equation 1.3) can be interpreted in terms of the fulfillment of the planned task of the center.

Let us consider how the center can control the time of consensus attainment m by the technical committee within the framework of the considered model to ensure inequality (Equation 1.3).

Let us introduce some additional definitions.

Let there be given a regular matrix $\mathbf{P}_1$ of size $n \times n$ that converges to the final matrix $\mathbf{F}_1$ after m_1 iterations. Convergence in this case is understood as the fulfillment of inequality

$$\det\mathbf{F}_1 \leq \varepsilon,$$

where ε is a given accuracy in the evaluation of the final matrix.

Let us call a regular matrix $\mathbf{P}_2$ of size $n \times n$ faster than the convergent one in comparison with a matrix $\mathbf{P}_1$ of the same size if the inequality $\det\mathbf{F}_2 \leq \varepsilon$ for the same value of ε is fulfilled through m_2 iterations and $m_1 > m_2$, where $\mathbf{F}_2$ is the final matrix for the matrix $\mathbf{P}_2$.

Let us associate the actions of the corresponding TC with the regular matrix $\mathbf{P}_1$, which actually describes the behavior of the members of this TC during the discussion of the draft standard. If the specific feature of the matrix

$\mathbf{P}_1$ is that $m_1 > \mathrm{M}$ (the planned time for discussion of the draft standard is violated), then in this case the reasonable strategy of the center is to arrange the transition to the more convergent matrix $\mathbf{P}_2$, for which $m_2 < \mathrm{M}$. Such a transition from the matrix $\mathbf{P}_1$ to the matrix $\mathbf{P}_2$ can be realized by introducing an additive control – such matrix **U** that the following condition is satisfied:

$$\mathbf{P_2} = \mathbf{P_1} + \mathbf{U}. \tag{1.4}$$

For the elements of the matrix **U**, certain conditions must be satisfied that ensure the regularity of matrix $\mathbf{P}_2$:

$$\forall i, j = \overline{1, n} \quad 0 < p_{ij}^1 + u_{ij} < 1, \ \text{i.e.} \ - p_{ij}^1 < u_{ij} < 1 - p_{ij}^1,$$
$$\forall i = \overline{1, n} \sum_{j=1}^{n} \left(p_{ij}^1 + u_{ij} \right) = 1, \ \text{i.e.} \ \sum_{j=1}^{n} (u_{ij}) = 0,$$

where $p_{ij}^1, \ u_{ij}$ are the elements of matrices $\mathbf{P_1}$ and **U**, respectively.

Since the increase in the level of authoritarianism of experts adversely affects the time to reach consensus, and the levels of authoritarianism (elements p_{ii}) are located on the main diagonal of the trust matrix $\mathbf{P_1}$, then any matrix $\mathbf{P_2}$, which is formed from matrix $\mathbf{P_1}$ by decreasing the value of the element $p^0{}_{ii} = \max\{p_{ii}\}$ with subsequent renormalization of the elements of the *i*-th row of the matrix $\mathbf{P_1}$, can be considered more convergent than $\mathbf{P_1}$. We call this procedure "weeding" the main diagonal of the matrix $\mathbf{P_1}$.

It should be noted that the "weeding" of the main diagonal of matrix $\mathbf{P_1}$ can be interpreted as a transition to the decision-making by the method of consensus minus one, consensus minus two, etc., which is used in the practice of some TCs during the development of "incomplete consensus" documents.

It may turn out that the value m_2 calculated for matrix $\mathbf{P_2}$ does not ensure the fulfillment of inequality (Equation 1.3). In this case, from the point of view of ensuring the inequality (Equation 1.3), a new element $p^1{}_{ii} = \max\{p_{ii}\}$ on the main diagonal of the matrix $\mathbf{P_2}$ should be chosen with subsequent renormalization of the elements of the *i*-th row of matrix $\mathbf{P_2}$, which is more convergent in comparison with matrix $\mathbf{P_2}$, etc. That is the meaning of management – to optimize or reduce the number of iterations in order to ensure inequality (Equation 1.3). In practice, this involves replacing one or more TC members. The decision to optimize or reduce the number of iterations (the replacement of a certain number of experts) should be made taking into account the human factor.

Moreover, it can be shown that, due to the choice of the control **U**, mathematically it is possible to obtain at one time the trust matrix $\mathbf{P_2}$ equal to the final one, i.e. to provide a consensus at once. For this it is sufficient, for example, to introduce a control in the form

$$\mathbf{U}= \begin{pmatrix} 0 & 0 & 0 & \dots & 0 \\ \Delta_{21} & \Delta_{22} & \Delta_{23} & \dots & \Delta_{2n} \\ \Delta_{31} & \Delta_{32} & \Delta_{33} & \dots & \Delta_{3n} \\ \dots & \dots & \dots & \dots & \dots \\ \Delta_{n1} & \Delta_{n2} & \Delta_{n3} & \dots & \Delta_{nn} \end{pmatrix},$$

where $\Delta_{ij} = p_{1j} - p_{ij} \forall i = \overline{2,n,}j = \overline{1,n.}$

Then in matrix $\mathbf{P_2} = \mathbf{P_1} + \mathbf{U}$ all rows are equal to the first row of matrix $\mathbf{P_1}$, which means, matrix $\mathbf{P_2}$ is final.

The introduction of the management center, formally connected with the change of the trust matrix, actually means the change of TC members (several experts or all). In principle, the approach associated with the change of individual members of the group is recommended to be used in order to achieve consensus. Consensus theorists, for example [12], believe that in some cases it is useful to exclude the stubborn from the group for achieving consensus.

1.6 The Quality of Consensus

Let us formulate the requirement, which it is useful to present to a consensus decision. Consensus will be considered as high-quality one, if each of the members made the same "contribution" to the final decision, i.e. no one "did not pull focus." Within the consensus model (Equation 1.1), this condition means that each row in the final matrix **F** has the form

$$1/n; 1/n; \dots; 1/n.$$

From the multiplication property of symmetric matrices it follows that for this it is sufficient to have such symmetric trust matrix **P** that the levels of authoritarianism of all members of the group were the same ($p_{11} = p_{22} = \dots = p_{nn} = p_0$) and the levels of the trust to the other experts were equal to

$$P_{ij} = \frac{1 - p_0}{n - 1}, \quad i \neq j.$$

Intuitively, this is understandable: since all members of the group are equally authoritarian and at the same time trust each other equally, their final opinion is completely balanced. This aspect is also noted, for example, in the manual for the achievement of consensus.[4]

The second aspect of the consensus quality is associated with the introduction of the control **U** from the center and the transition from matrix $\mathbf{P_1}$ to the more convergent matrix $\mathbf{P_2}$. Since these matrices converge to different final matrices $\mathbf{F_1}$ and $\mathbf{F_2}$, the question naturally arises: how far are the consensus solutions in the first and second cases? The answer to this is given by the following statement.

Let the column vector $\mathbf{S}^{\mathrm{T}}(0) = (\mathrm{s_0}^1, \ldots, \mathrm{s_0}^n)^{\mathrm{T}}$ be given in which the elements of the vector are in the interval $\mathrm{s}_a^i \leq \mathrm{s}_0^i \leq \mathrm{S}_b^i$, then the elements of the column vector $\mathbf{S}^{\mathrm{T}}(1) = \mathbf{P}_{n\times n}\cdot \mathbf{S}^{\mathrm{T}}(0)$ are also in the same interval, that is, the following condition holds for good for them: $\mathrm{s}_a^i \leq \mathrm{s}_0^i \leq \mathrm{S}_b^i$. This statement follows from Theorem 4.1.3 [13], according to which the estimate is true

$$\mathrm{S}_1^i - \mathrm{s}_1^i \leq (1-2\varepsilon)\left(\mathrm{S}_b^i - \mathrm{s}_a^i\right),$$

where ε is the smallest element of the initial trust matrix $\mathbf{P}_{n\times n}$, S_1^i and s_1^i are the largest and smallest elements of $\mathbf{S}^{\mathrm{T}}(1)$, respectively. In this case, $\mathrm{S}_1^i \leq \mathrm{S}_b^i$ and $\mathrm{s}_1^i \geq \mathrm{s}_a^i$. Continuing the chain of inequalities, for the elements of the vector $\mathbf{S}^{\mathrm{T}}(m)$ we have

$$\mathrm{S}_m^i - \mathrm{s}_m^i \leq (1-2\varepsilon)^n \cdot \left(\mathrm{S}_b^i - \mathrm{s}_a^i\right) \leq \mathrm{S}_b^i - \mathrm{s}_a^i,$$

where $\mathrm{S}_m^i \leq \mathrm{S}_b^i$ and $\mathrm{s}_m^i \geq \mathrm{s}_a^i$, which proves the original assumption.

Since no conditions other than regularity were imposed on the trust matrix, it is obvious that for identical initial views of the group members, the consensus solutions for matrices $\mathbf{P_1}$ and $\mathbf{P_2}$ are in the same interval. This feature allows us to hope that in reality the consensus when introducing the management from the center will not significantly differ from the agreed opinion of group members that could be achieved without the introduction of management, but for a longer time of discussion.

1.7 Consensus-Building Model Description Based on Cellular Automata Methodology

Cellular automata were discovered in the 40s of the 20th century by Hungarian-American mathematician John von Neumann in order to create a machine that would be able to reproduce itself [14]. A cellular automaton is

[4] http://book.od.ua/i/rekomendatsii-dlya-prinyatiya-resheniy-po-metodu-konsensusa

a discrete dynamic system in which each cell obeys to an adjusted rule set and works automatically without human involvement. To start cellular automaton the initial states of all cells and the rules of state transition should be set. The new state of each cell is determined at each iteration by using the rules of transition and the states of neighboring cells [15].

One of the most interesting and important points in the theory of cellular automata is that they could be used as an instrument of world cognition, for mathematics is unable to predict final state of some forms of cellular automata (it is obligatory to calculate every step of their evolution) [16]. Cellular automata are used in analysis of fingerprints, physico-chemical systems, different structure forms (seashells patterns, zebra stripes etc); they are also applied to modeling of crowd behavior, behavior of ants, bees, and wildfires [15]. Mechanisms proposed by Neumann for obtaining self-reproducing structures resemble the ones discovered in the next decade for biological systems [17].

The mathematical theory of cellular automata (aimed to investigation of properties of different systems and its classification) began to develop due to wide practical use in the 80s of the XX century [16]. Thus, cellular automata provide building useful models for many studies of natural and computational sciences. Thereby, one can search new approaches to old and yet unsolved scientific problems.

Let us build and investigate the achievement of consensus model based on cellular automata. To do this, describe the model, the initial conditions, and the rules of cell transition.

The first stage. Basic definitions and assumptions
Consider the opinions of members of some social group and represent them by a 2D square grid (Figure 1.4).

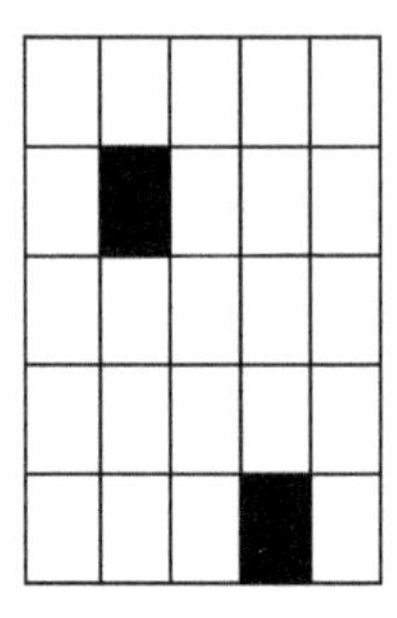

Figure 1.4 Space of opinions represented by a square grid 5×5.

We refer to this grid as space of opinions of the social group considered. Each cell of this grid set the opinions of one or several group members on the question under consideration. Grid dimensions define the area of possible opinions: the bigger the grid, the wider the area of various opinions.

The opinion of a group member could be presented in both short and detailed form. Quantity of neighboring cells could characterize this difference. This quantity we refer to as *number of degrees of freedom* for the cell considered. Depending on a cell position, this number of degrees of freedom may be equals to 8 or, 5 or, 3. We distinguish between three cell types:

- if the cell has the maximal number of degrees of freedom (8), then we refer to it as *central* one (Figure 1.5).
- if the cell has the minimal number of degrees of freedom (Equation 1.3), then we refer to it as *angular* one (Figure 1.6).
- if the cell has the medium number of degrees of freedom (Equation 1.5), then we refer to it as *boundary* one (Figure 1.7).

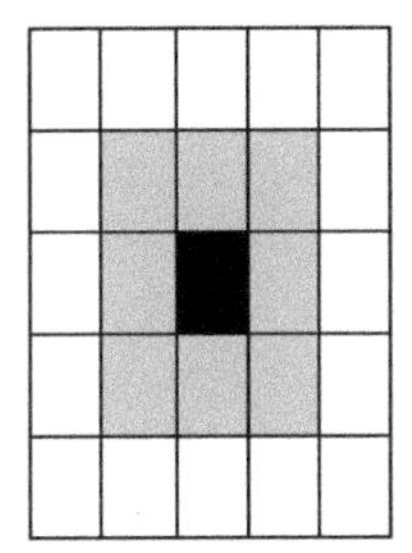

Figure 1.5 The opinion is set by *central* (black) cell. Gray ones indicate the number of degrees of freedom of the central cell (Moore neighborhood).

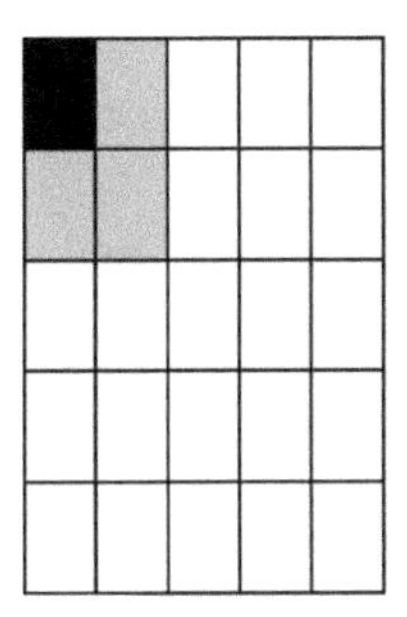

Figure 1.6 The opinion is set by *angular* (black) cell. Gray ones indicate the number of degrees of freedom of the angular cell.

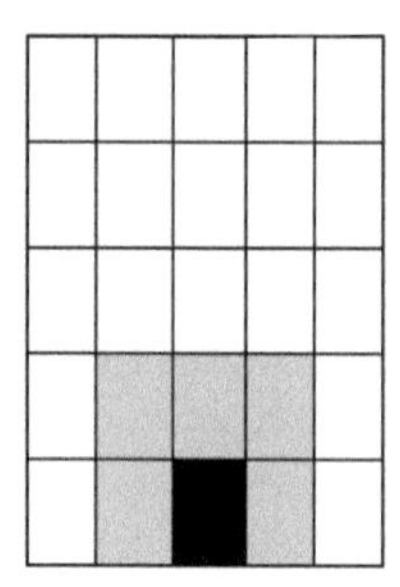

Figure 1.7 The opinion is set by *boundary* (black) cell. Gray ones indicate the number of degrees of freedom of the boundary cell.

Opinion is inextricably linked to one or several group members. In one agreement process (or iteration) each member could change his opinion, i.e. to agree with one or several participants. The opinion change is reflected by the cell position. Let us clarify this point:

- after one iteration the cell replaced another one of the same type. This means that one member has agreed with another, whose opinion is close to his own.
- after one iteration the cell replaced another one with a greater number of degrees of freedom. This means that one member has agreed with another, whose opinion was more detailed.
- after one iteration the cell replaced another one with a lower number of degrees of freedom. This means that one member has agreed with another, whose opinion was less detailed.

Thus, after one iteration, each group member may change his opinion or not.

The second stage. Initial conditions

Let us consider the agreement process from positions of two group members (of two negotiators). Let the members' opinions be different and each of them has its own cell (let us name these cells №1 and 2, respectively). Thereby, there are two moving cells in our cellular automaton.

The third stage. Transition rule

Let us set the transition rule for our cellular automaton: at one iteration each cell may stay on its own place or may randomly take a position of any neighboring cell (Figures 1.5–1.7). This means that at every agreement stage each member (or group) either does not change the opinion or not.

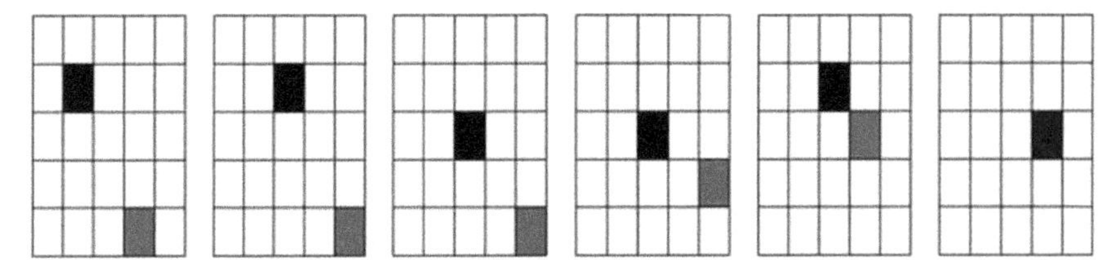

Figure 1.8 Work realization of cellular automaton.

In practice, this allows to determine the compromise/authoritarianism characteristic of the group member. If a member does not agree with any other at every agreement stage, then he can be considered as absolutely authoritarian member of the group. If a member agrees with any other at some agreement stages, then he is inclined to compromise.

The fourth stage. Work realization of cellular automaton

Let us define the notion of consensus at work of the cellular automaton built. Let the cells №1 and 2 take the same position in the grid after several agreement stages. This means that two initially different opinions of two members (or two groups) would match, i.e. *consensus* would be reached. We are interested in the number of transitions (iterations) leading to consensus, which characterize the quantity of agreement stages. This number sets *the time of consensus-building*.

Figure 1.8 shows the example of the work of cellular automaton with 5×5 grid until matching of the cells 1 and 2 positions, i.e. until consensus-building. As one can see, the cell 2 does not change its position after the 2nd and the 5th transitions, and the cell 1 does not do that after the 3rd transition. For a given initial cell positions and grid, the match occurred in five transitions, i.e. consensus reached in five stages of agreement.

1.8 Study of Consensus-Building Model Based on Cellular Automata Methodology

Let us study the factors influencing the time of consensus-building by means of the cellular automaton work analysis. These factors set the initial conditions of the cells 1 and 2 position and the grid dimensions. However, there are numerous combinations of the initial conditions and the grid dimensions we can set, so in order to reduce that number, let us apply the experiment planning [18], describe the stages of its realization, and analyze the work of cellular automaton under these conditions.

Assumptions and definitions

1. For modeling let us introduce the notion of distance between opinions and refer to it as Chebyshev distance between respective cells. Let there be two different cells, as in Figure 1.4, for example. The Chebyshev distance between them is $r = 3$. In practice, this distance could be presented as the minimal number of agreements steps before the consensus-building, providing the first member is inclined to compromise (according to assumption 1) and the second is absolutely authoritarian.
2. Let us introduce the notion of *the cell position change rate* at each transition:
   - if the cell does not change position, we assume its rate to be $\vartheta = 0$ at this transition;
   - if the cell changes position to one of the neighboring cells, we assume its rate to be $\vartheta = 1$ at this transition;
   - if the cell changes position to several cells, we assume its rate to be $\vartheta = r$ at this transition, here r is the distance between cells, defined in para. 1.

   In practice, the cell rate is related to *opinion instability* of the group member: the higher the rate, the more unstable the opinion of the group member and the more his tendency to cardinal change of his primary opinion.
3. For modeling let us introduce the notion of distance between opinions and refer to it as Chebyshev distance between respective cells. Let there be two different cells, as in Figure 1.4, for example. The Chebyshev distance between them is $r = 3$. In practice, this distance could be presented as the minimal number of agreements before the consensus-building, providing the first member is inclined to compromise (according to assumption 1) and the second is absolutely authoritarian.
4. We assume that the second member is inclined to compromise and in each agreement stage $\vartheta = 1$ (see para. 6). If for both participants in each agreement $\vartheta_1 = \vartheta_2 = 0$, then we come to the situation of having two absolutely authoritarian members in the group, and consensus will be unattainable (see para. 3.2 of the first approach).

The first stage. The response choice

As response let us consider the average duration of agreements until the consensus-building, i.e. the average number of the agreement stages conducted or the time of reaching a consensus.

The second stage. The factors choice

There are many ways for choosing factor combinations which are responsible for a variety of two cell position and their transitions. After a preliminary analysis was done, the following factors and levels were chosen:

X_1 – grid dimension *n:* 5×5, 11×11, 17×17;

X_2 – cell №1 position in the grid: the central, the angular, or the boundary one;

X_3 – cell 1 position change rate;

X_4 – distance between the cells *r:* *r* = min{the first member; the second member}; *r* = 0, 5{the first member; the second member}; *r* = max{the first member; the second member}.

The third stage. The factor-levels choice

A preliminary analysis of each factor's influence had shown that we are dealing with a non-linear model. For there was no assumption to go out of the frames of the quadratic polynomial, two or three factor-levels are enough in case of such model types. The choice of these factor-levels was made after the researches done. The results are given in Table 1.1.

The fourth stage. The planning matrix choice

A preliminary analysis had shown that the model would not be linear and there is a need to include a pair interactions and/or degrees. In order to do this, a full-factor experiment (orthogonal plan included 54 lines and 5 columns – 4 factors in total) was carried out. In a full-factor experiment all factor-level combinations are realized, which allows to estimate independently the influence of each factor and each pair interaction.

Table 1.1 Factors and their levels

Factors	X_1	X_2	X_3	X_4
Lower «–»	5	Mid central cell	0	min{the 1st member; the 2nd member}
Medium «0»	11	Mid boundary cell		max{the 1st member; the 2nd member}
Upper «+»	17	Angular cell	1	max{the 1st member; the 2nd member}

The fifth stage. The experiment

The plan includes a modeling with 1000 parallel experiments in each line. Each experiment (that corresponds to a line) includes the response information. In addition, the parallel experiments insure us from errors and allow us to estimate the error of reproducibility where there is a danger of having an experimental error.

For an imitating model the program in Python language was created. The modeling time was from 0.01 seconds to 30 minutes depending on the initial conditions. The analysis of results was carried out with the STATISTICA 7.0 software. For each parallel experiment the time needed for the second member to agree with the first one (consensus-building time) is estimated.

For each line the average duration of the game is calculated and the homogenity of dispersion is checked by Cochran's Q test for significance level of 0.01. Based on the results of the confirmed heterogenity at the chosen level of significance and constructed response histograms in each line of the plan, it was decided to consider the logarithms of the duration of the agreement stages, for which dispersion homogenity was confirmed at the level of 0.01 by Cochran's Q test [18].

The sixth stage. Calculations and equation

Preliminary, basing on a median range sweep and number of the points selected it is reasonable to include in the regression model the following factors: the grid size X_1, the distance between the cells X_4 and the pair interactions X_1X_4, X_2X_4 [18]. Use of the step-by-step inclusion of variables confirmed the inclusion of these factors and the following regression model was obtained:

$$\mathrm{LnY} = 3.15+1.07X_1+1.20X_4+0.62X_1X_4+0.31X_2X_4-0.32X_3X_4. \quad (1.5)$$

The plan matrix is orthogonal due to the full-factor experiment carried out, which allows us to interpret the model coefficients straightly. The model (Equation 1.5) is adequate at a significance level of 0,01 according to Fisher's criterion, the determination showing is high enough what also demonstrates the good quality of the model (Equation 1.5):

$$R^2 = 0{,}83;\ R^2_{adj} = 0{,}81.$$

The seventh stage. Results

The connection with the X_1 factor is the mostly evident in Equation (1.5): the space of opinions increase delays the consensus-building, other conditions being equal.

The graph of the model obtained is defined in a 5D space. Let us regard the sections of different levels in order to study this model deeper and interpret it. We start from the X_4 factor levels because for each level fixed we get the linear models.

1. Let us fix the X_4 factor on the upper level:

$$X_4 = +1; \; \text{LnY} = 4.35 + 1.69X_1 + 0.31X_2 - 0.32X_3. \qquad (1.6)$$

Now let us regard different response levels, counting that LnY varies from 1 to 7 in the experiment. In Figure 1.9 the sections of levels Ln2 (left section) and Ln4 (right section) are given. Because of model's (Equation 1.6) linearity it is easy to find the gradient $\text{Grad}_1 = \{1.69; 0.31; -0.32\}$ and then to analyze the factor-levels combinations for the most increase of LnY (and, hence, of Y itself). Any set of three coordinates that is proportional to the Grad_1 obtained also lies within this gradient. So we only have to choose the factor-levels

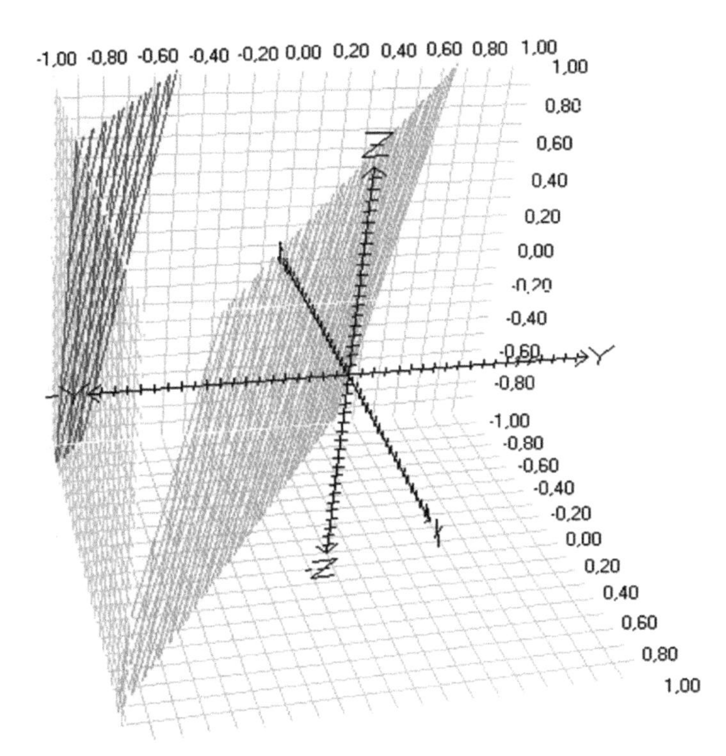

Figure 1.9 The 3D cube (experiment area) with surfaces of equal response values of Ln2 (left section) and Ln4 (right section) with $X_4 = +1$, i.e. at the maximal opinion distance.

that are usable for interpretation. To do that we use the «Steps» line as it is shown in Table 1.2 (for convenient searching factor-levels to interpret the coefficients and the gradient coordinates are rounded off to the tenths).

From the last line of Table 1.2 we infer the increase of number of agreements (at the condition listed in para. 1 if factors are the following:

- the grid size is 33;
- cell №1 is the angular one, the number of degrees of freedom is 3;

$$-\vartheta_1 = 10.$$

In practice that means: if initially the members' opinions are cardinally diverse, the consensus-building process is maximally delayed in case of one absolutely authoritarian member.

Also the model (Equation 1.6) demonstrates particular cases of one of the factors change:

- the increase of X_3 by 1 unit implies an average decrease of Y response $e^{0.33} \sim 1.38$ times;
- the increase of X_1 by one unit implies an average decrease of Y response $e^{1.69} \sim 5.42$ times.

Extrapolating the results behind the experiment's area gives us the case of cardinally distant opinions:

- the increase of the opinion change rate by 1 unit implies an average decrease of number of agreement stages in $e^{0.32} \sim 1.38$ times;
- the increase of grid dimensions by 1 unit (that corresponds to the addition of $2n + 1$ opinions in the opinion space to given n^2 opinions) increases, averagely, the number of agreements stages in $e^{1.69} \sim 5.42$ times.

Table 1.2 Search of the model (Equation 1.6) gradient and factor-levels combinations, $X_4 = +1$

	X_1	X_2		X_3
Main level	11	Cell	1 is mid boundary	None
Variation domain	2	2		2
Lower level	5	Cell	1 is mid central	$\vartheta = 0$
Upper level	17	Cell	1 is angular	$\vartheta = 1$
Model Bj coefficients	1.7	0.3		–0.3
$Grad_{11}$	3.4	0.6		–0.6
Steps	1.1	0.2		–0.2
Experiment with the highest value	5.6	1		–1
Total values	33	Cell	1 is angular	$\vartheta = 0$

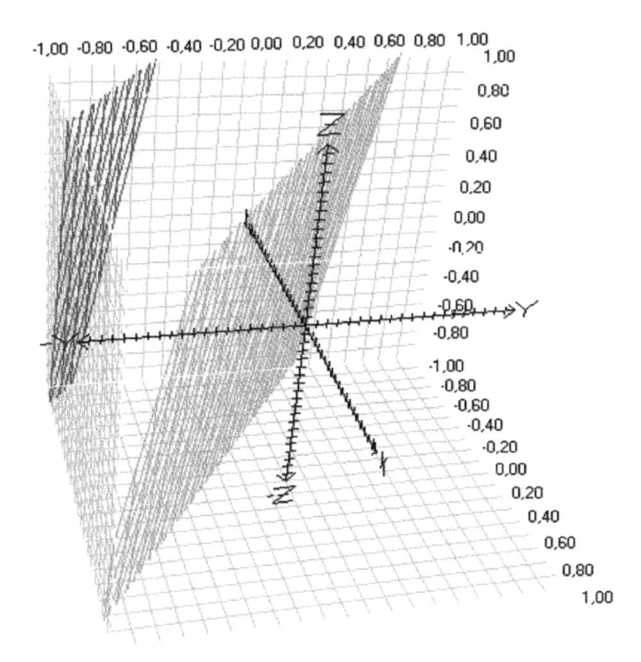

Figure 1.10 The 3D cube (experiment area) with surfaces of equal response values of Ln2 (left section) and Ln4 (right section) with $X_4 = -1$, i.e. at the minimal opinion distance.

2. Let us fix the X_4 factor on the lower lever:

$$X_4 = +1; \mathrm{LnY} = 1.95 + 0.45X_1 - 0.31X_2 + 0.32X_3. \qquad (1.7)$$

We obtained a linear model (as in para. 1) for LnY and it is easy to find the directions of the agreement stages' increase $\mathrm{Grad}_2 = \{0.45; -0.31; 0.32\}$ and corresponding values of factor-levels (Table 1.3). In Figure 1.10 the sections of levels Ln2 (left section) and Ln4 (right section) are presented.

From the last line of Table 1.2 we infer the increase of number of agreements (at the condition listed in para. 1 if factors are the following:

- the grid size is 21;
- cell №1 is the mid central one, the number of degrees of freedom is 8;

$$-\vartheta_1 = 1.$$

In practice that means: if initially the members' opinions are close, the consensus-building process is also maximally delayed in case of two absolutely compromise member (and their opinions are detailed). In other words, too loyal members will accept the parties of other members of the group, shifting responsibility to others, not trusting themselves.

Table 1.3 Search of the model (Equation 1.7) gradient and factor-level combinations, $X_4 = -1$

	X_1	X_2	X_3
Main level	11	Cell №1 is mid boundary	none
Variation domain	2	2	2
Lower lever	5	Cell 1 is mid central	$\vartheta = 0$
Upper level	17	Cell 1 is angular	$\vartheta = 1$
Model Bj coefficients	0,5	–0.3	0.3
$Grad_{21}$	1	–0.6	0.6
Steps	0,3	–0.2	0.2
Experiment with the highest value	1,6	–1	1
Total values	21	Cell 1 is mid central	$\vartheta = 1$

Also the model (Equation 1.7) demonstrates particular cases of one of the factors change:

- the increase of X_3 by 1 unit implies an average increase of Y response in $e^{0.32} \sim 1.38$ times;
- the increase of X_1 by one unit implies an average increase of Y response in $e^{0.45} \sim 1.57$ times.
  Extrapolating the results behind the experiment's area gives us the case of close opinions:
- the increase of the opinion change rate by 1 unit implies an average increase of number of agreement stages in $e^{0.32} \sim 1.38$ times;
- the increase of grid dimensions by 1 unit (that corresponds to the addition of $2n + 1$ opinions in the opinion space to given n^2 opinions) increases, averagely, the number of agreements stages in $e^{0.45} \sim 1.57$ times.

From paras. 1 and 2 we see that the maximal increase of agreement stages takes place in case of increasing the opinion space in combination with the remoteness of two members' opinions.

3. Continue to study the model: let us construct graphs of Y dependence at fixed X_3 and X_4 factor-levels (Figures 1.11–1.14). We are interested in particular conditions under which the number of agreement stages would grow along with the opinion space increase (X_1).

From Figures 1.13 and 1.14 we conclude that if the members' opinions are cardinally diverse ($X_4 = -1$), then the average number of agreement stages will be greater in case of the minimal number of degrees of freedom for the first member opinion (cell 1 is angular, see Figure 1.15). It is possible to say, that in this case of higher opinion mobility the consensus comes faster.

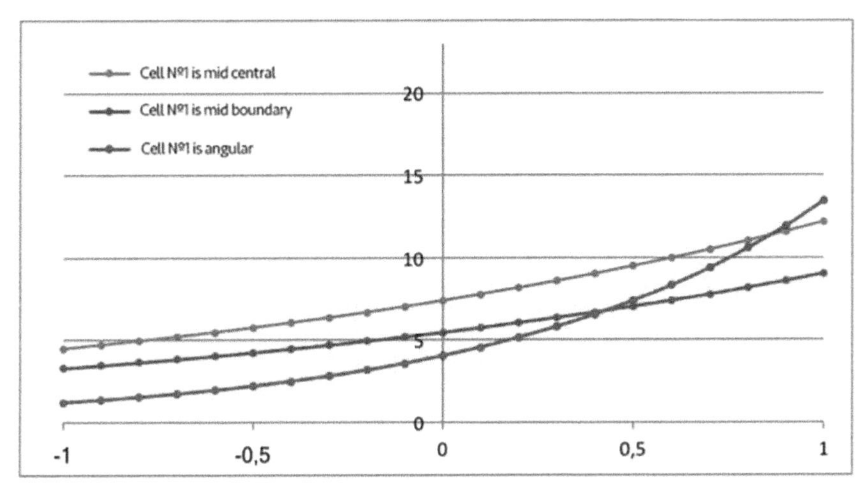

Figure 1.11 Dependency of the time of reaching a consensus Y on the dimension X_1 of space opinions: $X_3 = -1$, $X_4 = -1$ (i.e. the opinion of the first member does not change, the opinions are close).

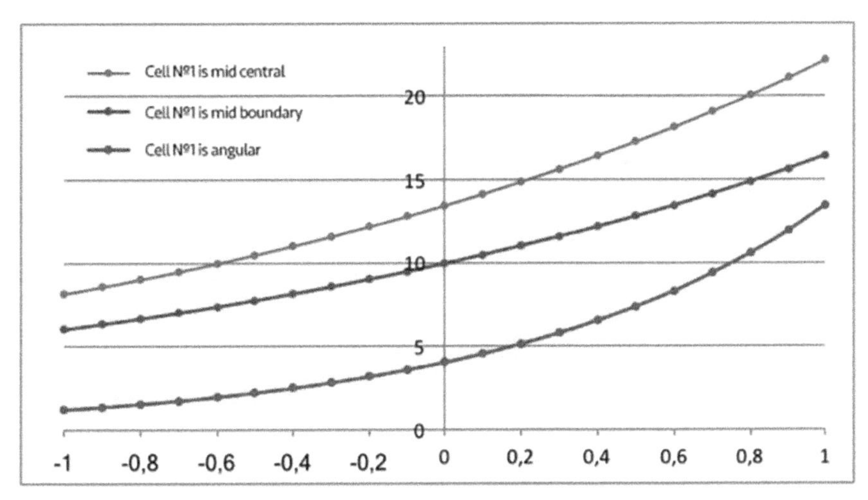

Figure 1.12 Dependency of the time of reaching a consensus Y on the dimension X_1 of space opinions: $X_3 = +1$, $X_4 = -1$ (i.e. the opinion of the first member changes, the opinions are close).

This conclusion agrees with the para. 1 conclusion of these results. As it was noted before, the authoritarianism of one of the group members negatively affects the time of consensus-building. On the other hand, the several times

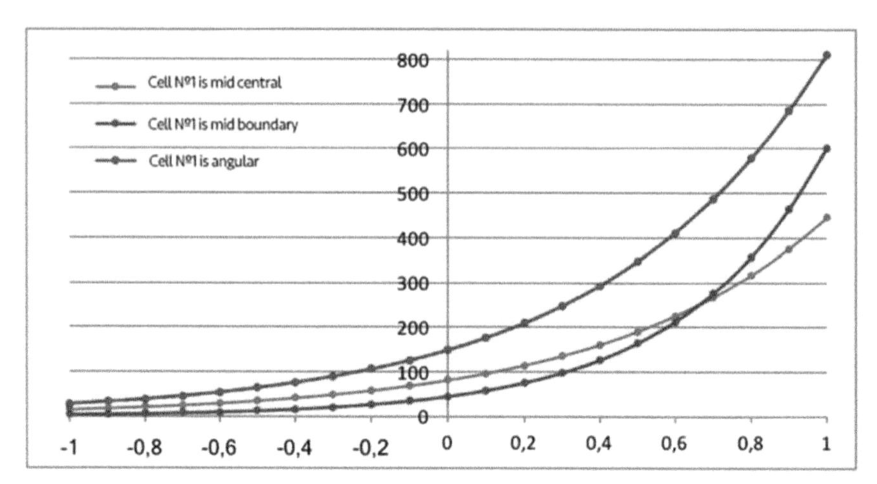

Figure 1.13 Dependency of the time of reaching a consensus Y on the dimension X_1 of space opinions: $X_3 = -1$, $X_4 = +1$ (i.e. the opinion of the first member does not change, the opinions are cardinally diverse).

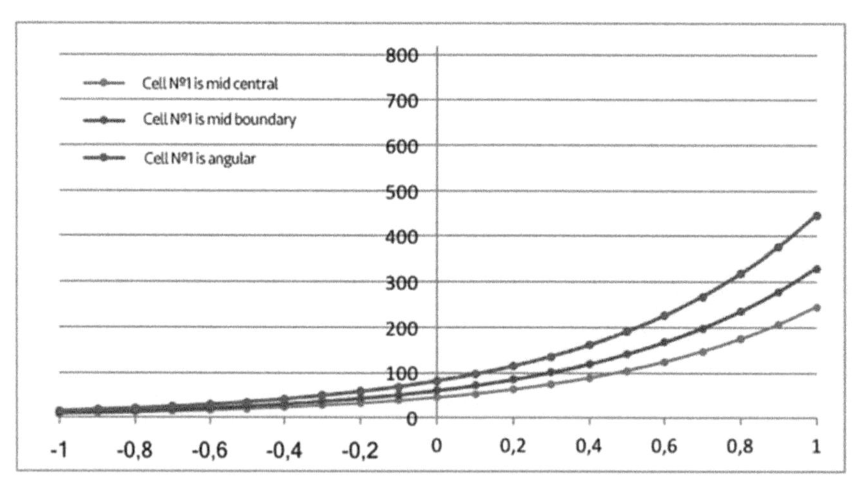

Figure 1.14 Dependency of the time of reaching a consensus Y on the dimension X_1 of space opinions: $X_3 = +1$, $X_4 = +1$ (i.e. the opinion of the first member changes, the opinions are cardinally diverse).

increase of number of agreement stages in case of opinion space and opinion difference growth should be pointed out.

From Figures 1.11 and 1.12 we conclude that if the members' opinions are close ($X_4 = -1$), then the average number of agreement stages will be

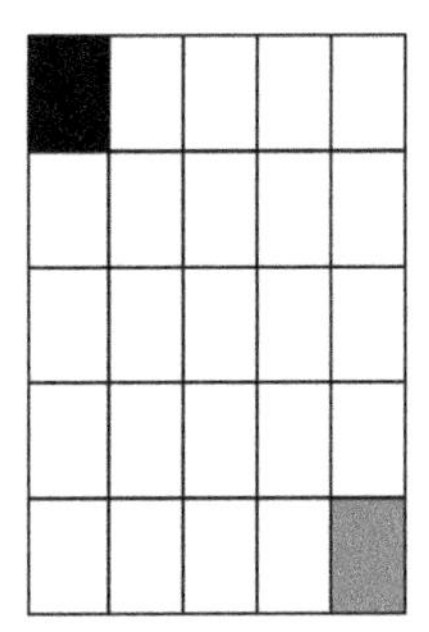

Figure 1.15 The black cell with the minimal number of degrees of freedom is the opinion of the first member, the gray cell at the maximal distance from the black one is the second member's opinion.

greater in case of the maximal number of degrees of freedom for the first member opinion (cell №1 is mid, see Figure 1.16).

If the members' opinions are maximally close ($X_4 = -1$), the average consensus-building time is longer in case of two members inclined to compromise and not in case of one authoritarian member. Really, in the second case the member inclined to compromise agrees with authoritarian member and consensus is reached faster.

Under fixed conditions on all factors, the opinion space increase (the grid dimensions increase) causes the number of agreement stages exponential growth (increasing curves on the Figures 1.11–1.14).

The number of agreement stages needed for consensus-building, averagely, increases in orders of magnitude in case of cardinally diverse ($X_4 = +1$) initial opinions.

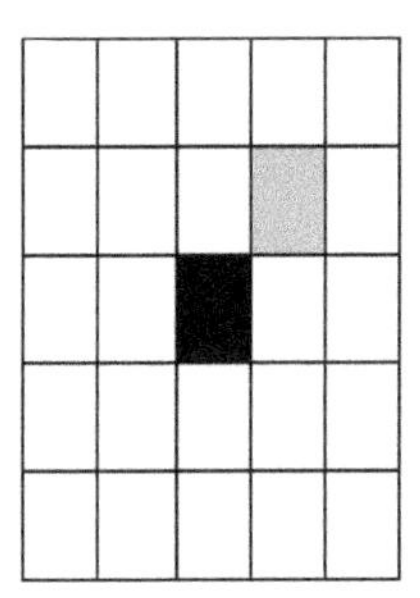

Figure 1.16 The black cell with the maximal number of degrees of freedom is the opinion of the first member, the gray cell at the minimal distance from the black one is the second member's opinion.

1.9 Conclusions and Results Interpretation

I. An important distinctive feature of the first model is the fact that it considers the consensus-building time dependency on the number of the group and the authoritarianism of its members. Also in the first approach a control is introduced, due to which it is possible to correct (to reduce) the number of agreement stages needed to consensus-building. This approach to consensus-building time analysis resulted in the following conclusions:

1. The increase of the number of the group, the TC in particular, the same as the increase of authoritarianism of the group members, negatively affects the consensus-building time and, hence, the work effectivity of the group.
2. For achieving consensus-building for a planned time control can be introduced. This control is to exclude situations (global dominance, presence of several leaders, responsibility shift, coalitions), in which consensus is fundamentally unattainable it will take significant time to attain consensus.
3. It is worth noting that the result obtained can be extended on any organization structures, in which decision is made on the basis of consensus: the increase of the number of the group in these structures and the increase of authoritarianism of its members significantly complicate the consensus achievement.
4. “Weeding” of the $\mathbf{P_1}$ matrix main diagonal can be interpreted as a transition to a decision-making by the method of consensus minus one, consensus minus two etc., which in practice is used in the development of “incomplete consensus” documents.

II. An important distinctive feature of the second model is the fact that it analyzes such characteristics as opinion space in the group and its structure (with a greater or lower number of degrees of freedom). The second approach to the consensus-building time analysis gave us the following results and conclusions:

1. A group disunity in opinions (the number of possible opinions) causes in average the exponential growth of agreements until consensus-building. The maximal increase of the number of agreement stages is found at the maximal opinion distance. A group disunity could be related both to the number of the group and to the complexity of the question regarded. This complexity should lead to the division of the question into smaller parts

in order to improve the quality of the agreements stages and minimize their number.

2. The opinion structure affects the consensus-building time. The agreement process, averagely, delayed in case of one authoritarian member with an uncontested opinion rather than in case of one authoritarian member with detailed opinion. In practice it means that it is much more difficult to find common points of contact with such a person in a question posed. This is consistent in a well-known fact in sociology (see for example [19]).
3. The number of agreement stages until consensus is achieved increases in orders of magnitude in a case of opinions' polarity. But if group members are inclined to compromise, then consensus achievement, averagely, occurs faster.
4. The initial opinions' closeness of discussion participants is not a guarantee of fast consensus-building if both participants do not maintain the constancy of their opinions. However, if in discussion one of the participants is firm in his opinion, then consensus is reached more quickly. Indirectly, this fact confirms the conclusions of Latane and Wolf [20] about the effect of firmness on the dynamics of processes in the social group.

III. An important methodological question arises: can the results of modeling really describe the behavior of people in the negotiation process with their variety of attitudes, characters, motivations, emotions, knowledge of the subject of negotiations, etc.?

It seems to us that a positive answer to the question posed is related to the fact that typical behavior of people in typical circumstances can be described in the same terms, which allows neglecting (in modeling) the listed factors that are difficult to formalize.

The following facts testify to the fact that people act in exactly the same way. In the book [15] it is indicated that a bag with $2 million dropped out of the truck that carried money. Drivers of cars passing by began to collect this money. As a result, the unfortunate driver was returned 200 thousand dollars (that is, one-tenth).

The Russian media reported[5] that on Thursday, June 8, 2017, a motorcyclist lost about 12 million rubles on a highway in St. Petersburg. Money fell from the man's backpack and flew along the road. He was returned about 2 million rubles (that is, one-sixth).

---

[5]https://lenta.ru/news/2017/06/08/12lyamov/

This circumstance allows us to hope that mathematical modeling nevertheless makes it possible at a qualitative level to correctly describe the behavior of people in interaction in a group.

References

[1] Della Porta, D., Andretta, M., Mosca, L., and Reiter, H. (2005). Globalization from below: transnational activities and protest networks. *Soc. Mov. Protest Content*. 26, 302.
[2] DeGroot, M. H. (1974). Reaching a consensus. *J. Am. Stat. Assoc*. 69, 118–121.
[3] Gantmacher, F. R. (1959). *The Theory of Matrix*. New York, NY: Chelsea Publishing Company, 276.
[4] Chebotarev, P. (2010). Comments on “Consensus and cooperation in networked multi-agent systems” *Proc. IEEE* 98, 1353–1354.
[5] Mazalov, V., and Tokareva, Y. (2012). Arbitration procedures with multiple arbitrators. *Eur. J. Oper. Res*. 217, 198–203.
[6] Klimek, P., Rudolf, H., and Thurner, S. (2009). To How Many Politicians Should Government to be Left? *Physica A: Statistical Mechanics and its Applications*, 388, 3939–3947.
[7] Anesi, C. (1997). The F-scale: Final Form. Available at: http://www.anesi.com/fscale.htm/
[8] Myers, D. (2010). *Social Psychology,* 10th Edn. New York, NY: MCGraw-Hill Companies, Inc., 759.
[9] Zazhigalkin, A. V., Aronov, I. Z., and Maksimova, O. V. (2014). Method mathematical model of consensus building in work of technical committees on standardization. *CDQM, Int. J*. 17, 48–55.
[10] Rogov, E. I. (2005). *Group Psychology*, Moscow, Vlados, p.430.
[11] Efron, B., and Tibshirani, R. (1991). Statistical data analysis in the computer age. *Sci. New Ser.* 253, 390–395.
[12] Gelderloos, P. (2006). Consensus: a new handbook for grassroots social, political, and environmental groups. Tucson, AZ: See Sharp Press.
[13] Kemeni, J. G., and S. J. Laurie. (1960). *Finite Markov Chains*. D. London: van Nostrand Co. Ltd., 210.
[14] vonNeumann, J., and Burks, A. W. (1966). *Theory of Self-Reproducing Automata*. Urbana, IL: University of Illinois Press, 388.
[15] Margolus, N., and Toffoli, T. (1987). Cellular automata machine. *Complex Syst*. 1, 967–993.
[16] Wolfram S. (2002). *A new kind of science*. Wolfram media, Inc., p.1197

[17] Gardner, M. (1970). The fantastic combinations of John Conway's new solitaire game "Life". *Sci. Am.* 223, 120–123.
[18] Fisher, R. A. (1966). The design of experiments/by Sir Ronald Aylmer Fisher-Edinburgh. London: Oliver & Boyd, 248.
[19] Moscovici, S. (1976). Social Influence and Social Change. New York, NY: Academic Press, 248.
[20] Latane, B., and Wolf, S. (1981). The social impact of majorities and minorities. *Psychol. Rev.* 88, 438–453.

2

Classification and Modeling of Intersystem Accidents in Critical Infrastructure Systems

Valery V. Lesnykh[1], Vladislav S. Petrov[2] and Tatiana B. Timofeeva[3]

[1]The National Research University Higher School of Economics, Energy Institute, Moscow, Russia
[2]NIIgazeconomika, LLC, Moscow, Russia
[3]The State University of Management, Moscow, Russia

The chapter is devoted to the problems of classification and quantitative assessment of intersystem accident consequences including cascade failure process. Classification of intersystem accident is proposed based on topology of cascade process. Topology-based and flow-based approaches are used for modeling of intersystem accident in power and gas supply systems. Some approaches for perturbance propagation description are under discussion.

2.1 Introduction

The urbanization process is growing rapidly during the latest several dozens of years. According to Ref. [1] 30% of world's population was urban in 1950, 54% in 2014 and it is expected to reach 66% in 2050. The most urbanized regions include North America (82%), Latin America (80%) and Europe (73%). The urbanization leads to mega-cities rising, by 2030 the world is projected to have 41 mega-cities with more than 10 million inhabitants. At the same time the number and scales of city and industrial agglomerations increase. The processes listed above lead to proliferation of infrastructurally-complex territories—territory with high concentration and high level of interaction of infrastructural systems (Moscow and the Moscow region, Düsseldorf-Cologne area, etc.). Important feature of infrastructurally-complex territories is high density of critical infrastructures.

The critical infrastructures comprise such main life support facilities as power supply (electricity, natural gas, petroleum products, heat), transport, water supply and water disposal, telecommunications, etc. The above listed systems are interrelated by material, power engineering, and informational flows.

Meant by the intersystem failure (ISF) will be such a development of abnormal processes, when the initiating event in one system leads to negative consequences (equipment breakage, collapse of buildings and structures, inventory losses, damage to health or loss of life, deterioration in environmental quality, etc.) in other interrelated systems. Meant by the ISF risk will be the anticipated summarized negative implications caused by intersystem development (including the cascading one) of the abnormal processes. One of the major tasks in ensuring safety and stability of interrelated critical infrastructures consists in identifying the places where propagation of disturbance among the systems is possible. The topology of propagation of disturbance among the interacting systems and within each system has been studied not as comprehensive as needed for the time being.

Simulation of cascading abnormal processes in separate critical infrastructures should be considered to be one of the first steps in investigating the intersystem failures. The cascading failures are investigated to the fullest extent possible pertinent to electrical power systems. For several dozens of years the problems of simulating and managing the cascading failures were studied by a number of researches. A considerable number of results on this problem are presented in [2, 3]. Special attention has been paid to the issues of formation of mechanisms for ensuring reliability of the electrical power systems. At the same time, this publication has not considered the issues of simulation of the cascading intersystem failures, as well as the respective risk assessment.

The publications of the group of authors from Refs. [4–6] are another example of comprehensive analysis of the cascading failures in the electrical power systems. These publications have developed the model of statistically-distributed branching Galton–Watson processes pertinent to the cascading failures in the electrical power systems. The authors use a classical definition of the risk of the cascading failures, which is understood as the product of probability (frequency) of the cascading failures leading to power supply interruptions by the damage caused by interrupted electrical supply. The same electrical supply has made an attempt to expand the used models for description of the cascading failure in two interrelated critical infrastructures [7].

Noticeably lesser studies have been devoted to simulation of the cascading failures in the gas supply [8, 9] and heat supply [10] systems. Simple model approach for cascade accidents in transportation and telecommunication systems is presented in [11].

The problem of risk assessment in interrelated critical infrastructures has been most comprehensively considered in the monograph [12]. Notwithstanding the fact that the articles included in the monograph are of a methodological and qualitative nature, thus far they present the fullest treatment of the range of problems pertinent to interacting critical infrastructures. In particular, the publication has suggested the rating of interrelations (physical, informational, geographical, logical), analyzed the models for risk assessment with allowance for interrelations of the systems, as well as analyzed the statistical data on failures in interacting critical infrastructures. It should be noted that in this publication the risk assessment in interrelated systems was to a large extent targeted to vulnerability analysis. At the same time, the quantitative risk assessment has been made in this publication only for two interrelated infrastructures.

Latest investigations connected with critical infrastructure and cascading accidents were focused on term of their roots and triggered to events in relation to society's feedback loops, rather than nature [13].

It is worthwhile to say that several software products making it possible to assess the consequences of the intersystem failures have been developed by now, for instance [14]. A 3D simulation model for emergency interaction of major critical systems (power supply, gas supply, heat supply, water supply, ground transport) for the city of Berlin has been implemented within the framework of this project. The SIMKAS-3D C model enables to reveal the places of physical concentration of the infrastructural systems and the abnormal process, including the damage assessment, is simulated for these places. On the whole, this model makes it possible to carry out the risk assessment in the intersystem failures, but this is true only for the cases of physical effect of the infrastructural systems.

The methodological framework of studying the risk of the intersystem failures should be referred to the notion of "system of systems" [15]. Further methodological development of this range of problems, to our opinion, takes place within the framework of the notion of "resilience". Resilience as a comprehensive methodology also comprises such interdisciplinary researches as risk assessment and management, provision of security and protection of the critical systems, as well as prevention of failures and catastrophes and elimination of their consequences [16].

The chapter focused on the two problems connected with intersystem accident risk assessment. First problem is connected with extension of existing intersystem failures classification. The preliminary analysis of the intersystem failures shows that the failures affecting two and more life support facilities are the most hazardous ones and often cause disastrous consequences. The intersystem failures substantially differ by the sources initiating the ISFs, scenarios of the abnormal process development, duration of exposure to negative implications, as well as the number and kind of the infrastructural systems involved and scale of subsequences. In this connection it is expedient to classify the intersystem failures, which will make it possible to substantiate the approaches to simulation of the abnormal processes in the interacting infrastructural systems.

Classification of intersystem failures types offered in Ref. [17] is limited to three types: common causes, cascading, and escalating. The analysis of the occurred intersystem failures, as well as the qualitative analysis of possible topologies of the ISF scenarios made it possible to suggest extended classification of the ISF structures focused on branch (cascade) processes.

Second problem is connected with modeling of intersystem accident including cascades. Various approaches to a quantitative estimation of risk of intersystem failures are under investigation. The fullest classification of used approaches is given in work [18] where the basic types of models concern: empirical, agent based, system dynamic based, economic theory based and network based. The most preferable from the point of view of a number of criteria (level of a readiness of methods, the account of all types of interdependence between systems, resilience level estimation, etc.) are network-based approaches. This approach includes topology-based and flow-based methods. The both methods have been used for quantitative analysis of intersystem accidents risk.

Within the problem of intersystem failures modeling a particular interest and difficulties are connected with the description of cascade failures. The most in detail given problem of research with reference to electricity supply systems (for example [6, 7]). Cascade development of failures also was investigated with reference to abstract interdependency systems [15].

Other problem is connected with an approach to the interaction between systems during failures description. Basically, the most general approach is connected with use of financial flow as general equivalent. Such approach, in particular is used in agent-based approach, for example in input–output inoperability model [19]. Within the proposed approach (network-based model) it is expedient to use a power equivalent for the interaction description.

2.2 Examples of Intersystem Failures

The preliminary analysis of ISF shows that the most dangerous failures often causing catastrophic consequences are the failures affecting two and more life-support systems. Examples of such failures are resulted in Table 2.1.

Table 2.1 Examples of the largest intersystem failures

Date, Country, Region	ISF Description
1977, USA, New York	Because of lightning hit in electric main lines New York city with suburbs remained without an electrical supply from July, 13th till July, 14th. Rainwestwood cogeneration plant switching-off at 21:27 led to stop of transport, mobile communication, public utilities.
1979, Russia, Norilsk	Rupture of gas pipeline occurred at temperature minus 50°C. As a result of sharp pressure drop the deforming wave extended on 58 km. As a result of failure more than 40 km of the pipeline were destroyed. The city of Norilsk remained without heating for some days. There were also considerable faults in an electrical supply, work of transport, municipal enterprises. Failure consequences were eliminated within three days. Cause damage cost was 2.8 million RUR.
2003 USA, Canada	On August, 14th in a number of the largest cities of east coast of the USA and Canada there was the technogenic accident which received the name «Black-out 2003». The electricity was disconnected in the cities of New York, Detroit, Cleveland, Toronto, Ottawa and others. Failure in power system led to fan power cutoffs in the area more than 24 thousand square kilometers. The emergency impacted more than 50 million persons, led to stop of over 100 power stations, including 2 nuclear power plants in both countries. On reapplication of power supply left more than a day. The damage because of global blackout in eight north-east states of the USA was more than $6 billion.
2003 Apennine Peninsula	Tree falling initiated a system collapse electrical supply system of Italy including Vatican and San Marino. There was a mass switching-off of consumers of the electric power. 57 million inhabitants spent without an electrical supply from 5 till 16 hours. Power cut off led to considerable failures in public utilities, transport, communication systems.
2004 North Korea	There was a large railway accident on April, 22nd at rail way station Ronchhon in 50 km from Pyongyang. Cars with ammonium nitrate were derailed and touched electric mains. The electric cable under a high voltage fell to the car with chemicals therefore there was an explosion. About 1750 houses were destroyed, about 300 persons were lost, more than 3000 got wounds.
2005 Russia, Moscow	On July, 25th the fire on Chaginsky electrosubstation of Moscow led to system collapse, the electricity supply in majority part of Moscow

(Continued)

Table 2.1 Continued

Date, Country, Region	ISF Description
	and Moscow suburbs was disconnected. There were failures in the power supplies at 50 stations of the Moscow underground, 30 water pump stations stopped—in a number of areas of Moscow and Moscow suburbs inhabitants remained without water. Work of the industrial enterprises, services, telecommunication transport was stopped.
2007 Ukraine, Vinnitsa Region	There was an explosion of gas to the further ignition on a gas pipeline of "Urengoj-Pomary-Uzhgorod". About 30 m of a gas pipeline was damaged, the site of a gas pipeline in length about 25 km was blocked. Failure led to the gas supply termination in 22 settlements of Vinnitsa area with the population about 36 thousand persons.
2008 Russia, Ulan-Ude	In February there was a major accident in cogeneration plant, all six coppers of the plant were disconnected, 524 houses of the city were without a heating—over 170 thousand inhabitants. In the city the emergency situation mode was entered, the electrical supply was limited, there were faults in transport, primary and secondary schools were closed.
2009 Russia, Krasnoyarsk Region	On August, 17th there was a failure on Sajano-Shushensk hydro power station located on the river Yenisei in Siberia. Failure occurred during repair of one of hydroturbine units. As a result of failure the third and fourth water channels were destroyed, there was a destruction of a wall and flooding of a machine hall. Nine of ten water-wheels completely failed, the station was stopped. Because of failure power supply of the Siberian regions was broken, fan switching-off concerned of some the industrial enterprises, including the Siberian aluminium factories. As a result of failure 75 persons were lost, 13 suffered.
2011 Mexica, USA	On September, 8th more than 10 million inhabitants of Mexico and the USA remained without electricity supply. Power cut off occurred in the Mexican cities of Tijuanas, Ensenada, Tekate, the Dignity Luis Rio Colorado and also in San Diego and a southern part of Los-Angeles. Two nuclear power plants were automatically disconnected in California. The reason of switching-off of electricity was the mistake of worker of electric company personal of Arizona Power Service (APS).
2012 India	On July, 31st in India there was an energy crisis which mentioned 22 states of the north, the west and the east of the country. The crisis reason was excess of norm of power consumption of four northern states Rajasthan, Haryana, Punjab and Utta-Pradesh. From faults with the electric power suffered more than 600 million persons. Northern, eastern and north—eastern parts of power system were worked with perturbation. Electricity delivery in Delhi fell from 4 thousand megawatt to 40 megawatt, all six lines of the Delhi underground were stopped. In the north of India more than 500 trains has been suspended. In state of West Bengal some hundreds miners were blocked underground.

2016 Australia	Widespread blackouts occurred beginning late on Tuesday December 27th, with areas losing power for upwards of twelve hours following severe storms causing damage to over 300 powerlines in the electricity distribution network. The storms also caused flooding and wind damage, including property destruction due to fallen trees. A total of 155,000 properties lost power at the peak of the storms, requiring over 1200 repair jobs resulting from over 350 powerlines being damaged. On Thursday December, 29th, there were more than 11 500 households still without power across the state, some for up to forty-six hours, in regions including the Adelaide Hills, Mid-North, Flinders Ranges, and Murraylands. By 9 am on Saturday December 31st, there were still more than 1600 households without power for more than 80 hours, primarily across the Adelaide Hills.

The occurred intersystem failures essentially differ with sources of initiation, scenarios of development of emergency process, duration of negative consequences display, number and a kind of the involved infrastructural systems and scale of consequences. In this connection it is expedient to classify of ISF that will allow to prove approaches to modeling of emergency processes of interconnected infrastructural systems.

2.3 Classification of Intersystem Failures

Let's view interacting of the several interconnected systems. Each of systems is structurally-complex and it is possible to present it in the form of bound guided or the undirected graph. Interacting between systems also is representable in a view of the graph where knots interreacting between systems are related. In each i-th knot of system r the peak permissible load C_i can be presented as:

$$C_i^r(t) = \alpha_i(t) L_i^r,$$

where L_i^r – loading (or short shipment, depending on system type) on a knot in the unperturbed system r, $\alpha_i(t) > 1$ – parameter, generally dependent on failure duration and size of perturbation which the given knot without loss of the full functionality is capable to withstand. The maximum load in knots of the count of intersystem interacting can be similarly presented:

$$C_k^{l-m}(t) = \beta_k(t) L_k^{l-m},$$

L_k^{l-m} – a loading/short shipment on a knot in the unperturbed operation of systems l and m, $\beta_k(t) > 1$ – parameter of specifying size of perturbation which the given knot without loss of the full functionality depending on a perturbation time is capable to withstand.

Let's guess that in one of systems in one of knots there is an initiating perturbing event (malfunction, accident) with duration t_1. In this case on system knots r and knots of intersystem interacting $\tilde{L}_i^r(t)$ and $\tilde{L}_k^{l-m}(t)$, $t \leq t_1$, accordingly. If during a time t_1 the perturbed loading in a system knot r $\tilde{L}_i^r(t)$ exceeds a maximum load ($\tilde{L}_i^r(t) > C_i^r$) or, analogously, in a knot of gateway interacting $\tilde{L}_k^{l-m}(t) > C_k^{l-m}(t)$, such knots we will consider disabled. Let mark $N^r(t)$ as number of disabled knots of a system r in a time moment t, N_{tot}^r – the full number of knots of a system r. Let define defeat of system r (damage level) as $\xi^r(t) = \frac{N^r(t)}{N_{tot}^r}$ accordingly, $N^{l-m}(t)$ – number of disabled knots of intersystem interacting l–m in a time t, N_{tot}^{l-m} – the full number of knots of intersystem interacting l–m, $\xi^{l-m}(t) = \frac{N^{l-m}(t)}{N_{tot}^{l-m}}$ – defeat of knots of intersystem connection. At inoperability of a knot during a time t_2 the loading on the remained knots is equal in system $\tilde{\tilde{L}}_i^r(t)$ where $t \subset t_1 \cap t_2$.

The peak values of parameters $\xi^r(t)$ and $\xi^{l-m}(t)$, in time $t \in (0, \max\{t_1 \cdots t_p\})$, allow to allocate four types of development of intersystem failures (Table 2.2).

The failures without branching (Type I) are the most frequent case of ISFs. Branching of the abnormal process in one of the interacting systems (Type II) leads to heavier consequences. Theoretically the ISF can develop in such a way that the intersystem cascade takes place (Type III), while the cascading processes may not be observed in each of the interacting systems. Besides, feedback of abnormal processes may theoretically occur among the interacting systems. It is necessary to underline that Type III needs statistical evidence. The heaviest consequences appear when branching of the abnormal processes in the systems and among the systems is realized (Type IV). Graphic representation of possible topology of scenarios of ISF development is shown on Figure 2.1.

Table 2.2 Classification of intersystem accident [20]

ξ^r	ξ^{1-m}	Type
$1/N_{tot}^r$	$1/N_{tot}^{1-m}$	I. Lack of branching
$1/N_{tot}^r < \xi^r \leq 1$	$1/N_{tot}^{1-m}$	II. Branching in systems
$1/N_{tot}^r$	$1/N_{tot}^{1-m} < \xi^{1-m} \leq 1$	III. Branching between systems
$1/N_{tot}^r < \xi^r \leq 1$	$1/N_{tot}^{1-m} < \xi^{1-m} \leq 1$	IV. Branching in and between systems

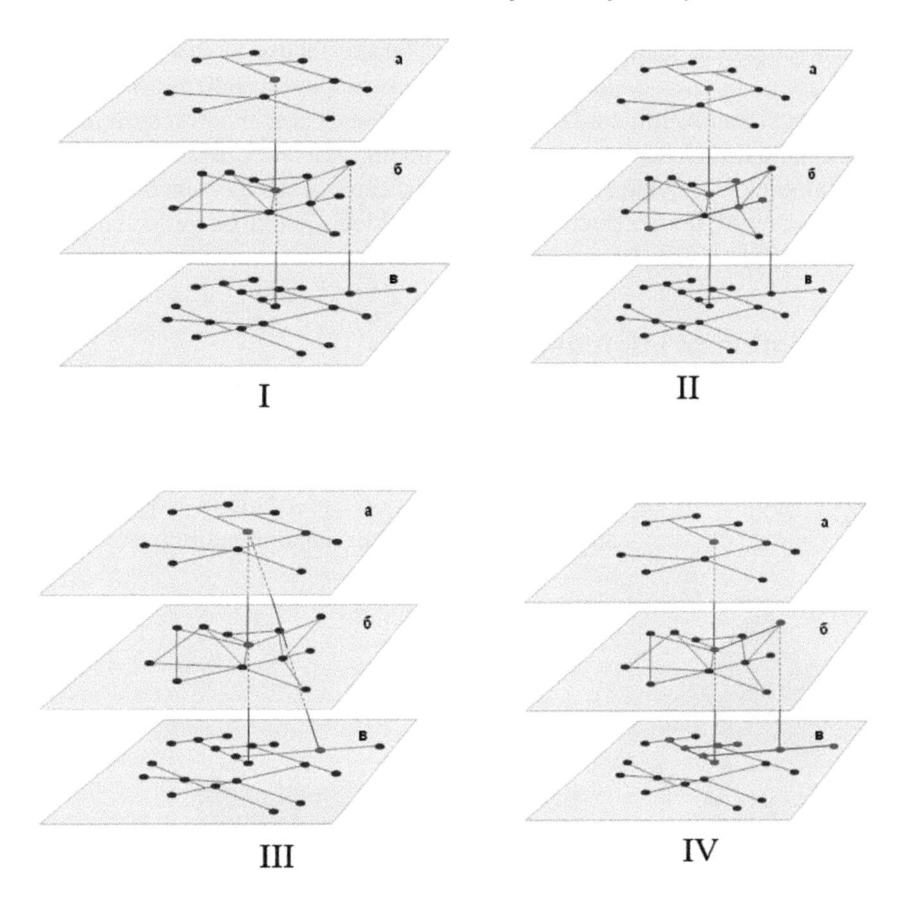

Figure 2.1 Classification of ISF by structure of emergency process development.

The analysis of the suggested classification by the structure of development of abnormal processes makes it possible to draw some conclusions. The failures without branching (Type 1) are the most frequent case of ISFs. Branching of the abnormal process in one of the interacting systems (Type 2) leads to heavier consequences and also takes place rather frequently. Theoretically the ISF can develop in such a way that the intersystem cascade (Type 3) takes place, while the cascading processes may not be observed in each of the interacting systems. Besides, feedback of abnormal processes may theoretically occur among the interacting systems. The heaviest consequences appear when branching of the abnormal processes in the systems and among the systems is realized (Type 4).

ISF development scenarios are initiated by the failure in the electrical power systems. Figures 2.2 and 2.3 show the example of realizing the ISF development scenarios initiated by the failure in the electrical power systems.

A substantial difference in the failure running scenarios is clearly seen on the example of analyzing ISF, including the type, sequence, and duration of involvement of the interacting systems, which should be reflected in simulation of ISFs.

2.4 Simulation of Intersystem Failures

The study was divided into several stages [20]. At the first stage the simple model of interaction of two diverse systems—electric and gas transmission networks was developed. The abstract gas-transport and electrical systems, topologically close to corresponding systems of Great Britain have used. Possible balance of energy streams of each system, under condition of their

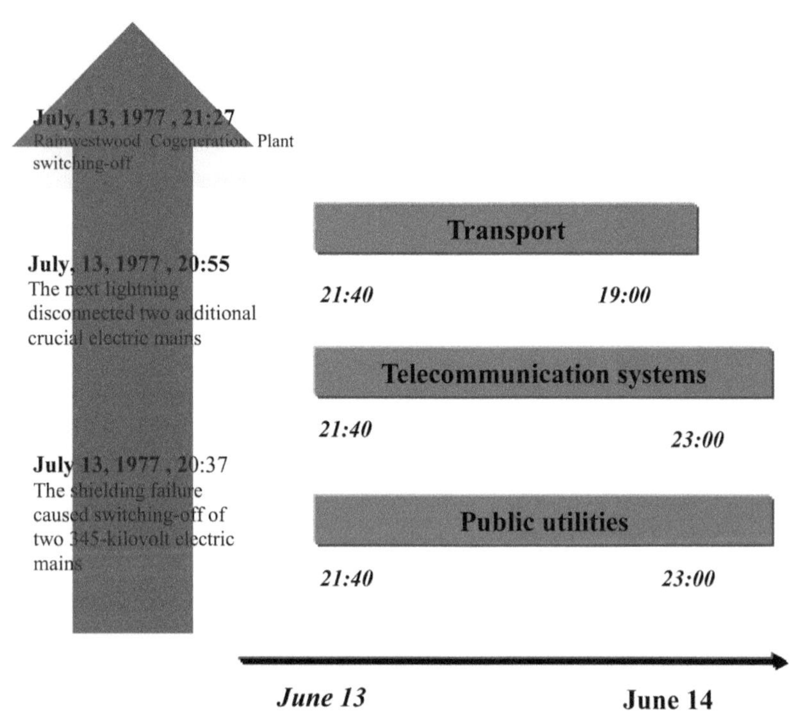

Figure 2.2 Diagram of ISF development in electric power system of the city of New York in 1997.

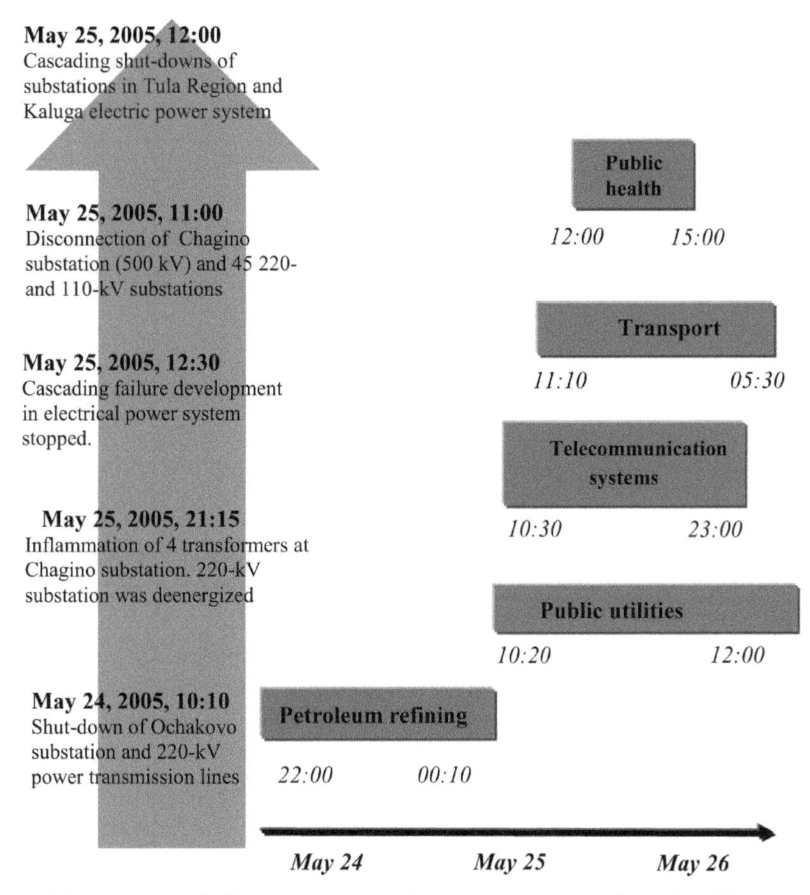

Figure 2.3 Diagram of ISF development in electric power system of the city of Moscow and Moscow Region in 2005 [20].

interacting at a diversion of an operating mode of one of system from the optimum was calculated. A possibility of use of underground storages of gas which allow to compensate originating diversions from an optimum operating mode of a gas-transport system did not consider.

2.4.1 Gas Transmission Network Model

In the gas transmission network model the main gas pipeline division is presented as a related directed graph: $G = (V, E)$ where V is the vertex (knot) set, E is the oriented edge set. Graph vertexes G (V) are the facilities of the gas transmission network, which is essentially either the gas source, or its drain, or the node where the flow value is varying (for instance, intake for own needs of the gas-compressor station). Assign the numbers to the network nodes in a gas flow direction. The network-oriented edges are the line sections of the gas transmission network. The net gas value in the node is determined as a difference between the incoming and outgoing flows. The node is considered to be a source if the net gas value is positive or a drain if the net gas value is negative. If node *i* is neither drain, nor source, the flow conservation concept (continuity equation) is true for it:

$$\sum_{j\in\alpha_i} Q_{ij} - \sum_{j\in\beta_i} Q_{ij} = 0$$

$$Q_{ij} \leq C_{ij}$$

Here Q_{ij} is the gas flow, β_i is the set of all nodes related to node *i* by means of the incoming oriented edges, α_i is the set of nodes related to node *i* by means of the outgoing oriented edges, and C_{ij} is the throughput capacity of the oriented edge.

Gas motion over the oriented edge between the nodes is described in terms of the system of one-dimensional gas-dynamic divergence equations, which are due to the conservation laws. Solution of this system makes it possible to determine the field of variable values p, ρ, v, T for the unsteady gas flow where p, ρ, v, T is the density, pressure, speed, and temperature of gas, respectively. However, while considering the steady-state gas flow and equation of state expressed in terms of $p/\rho = zRT$ where *z* is the gas non-ideality factor, the equations can be simplified and one can neglect the factors of the second order of smallness and determine the gas flow in the oriented edge:

$$Q_{ij} = 0.0385\sqrt{\frac{p_i^2 - p_j^2}{zT\lambda\Delta L_{ij}} D^5} \tag{2.1}$$

Here p_i, p_j is the initial and final pressures in the gas pipeline division, Δ is the specific gravity of gas L_{ij}, the oriented edge length, λ – coefficient of hydraulic resistance. Then in the *i*-th node with preset drain or source Q_i the pressure drop will amount to $\Delta p^2 = B_0 Q_i^2 \lambda L_{ii-1}/D^5$ where $B_0 = \frac{zT\Delta}{K^2}$.

If the *i*-th node is the gas-compressor station, its performance equation can be presented in terms of performance $p_{\text{Hi}}{}^2 = a_i p_{\text{Bi}}{}^2 + b_i Q^2$, где a_i и b_i are the trial coefficients depending on the gas composition (z, R) gas temperature at the inlet of blower T_{E} and revolutions per minute n [21]. Such approach at modeling of a gas-transport network allows to consider the effects connected with partial switching-off of power plants of compressor stations.

The system of equations determined in this manner makes it possible, with allowance for the network topology, to calculate its mode of operation and determine the values of each Q_{ij} in the unified gas-dynamic system. The system of equations determined in this manner makes it possible, with allowance for the network topology, to calculate its mode of operation and determine the values of each Q_{ij} in the unified gas-dynamic system. The unperturbed gas-transport system is characterised by the given balances of gas flows of each knot. In case of defeat of compressor station in a knot of a gas-transport system, the stream on the edges of a graph related to this station impinges approximately in 1.4 times as there is a necessity to adjust a pressure modification. The new operating mode of gas-transport system further was calculated. The requirement of a minimum diversion from the unperturbed operating mode was measure for a select of a new condition of operation.

Figure 2.4 shows the gas transmission network model used in this publication. Red circles stand for sources, blue circles, for drains, and dark blue circles, for gas-compressor stations. The assumption is made for permanence

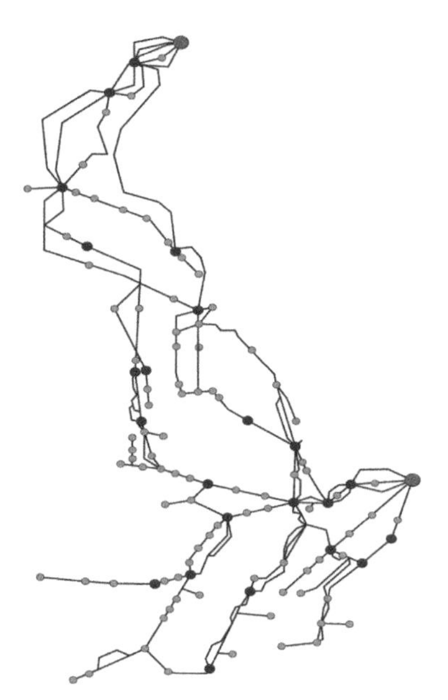

Figure 2.4 Gas transmission network model. Sources are shown with red circles. Blue circles are users. Dark blue circles are gas-compressor stations.

of the network topology, though in the general case the network topology may vary rather substantially depending on the operating procedures, scheduled repairs, modes of operation of the underground gas storage facilities, etc.

2.4.2 Electric Network Model

The high-voltage (300–400 kV) electric network model is essentially unoriented graph $\tilde{G}(V, E)$Vertexes $\tilde{G}(V)$ of the graph are either the power plants (sources), or distribution substations of given power w_i oriented edges $\tilde{G}(E)$ of the graph correspond to the power transmission lines with given efficiency ε_{ij}. Each node of the network is connected to any related power source over the $\min(\sum \varepsilon_{ij})$ shortest (minimum) route in the network. Load L_i at the *i*-th node of graph $\tilde{G}$ is determined as the number of minimum routes passing through this node multiplied by power w_j of final node *j* fed over this route [22, 23]. In each *i*-th node of graph $\tilde{G}$ maximum permissible load C_i is determined:

$$C_{i=}\alpha_i L_i$$

where L_i is the load upon the node in the unperturbed network, $\alpha_i > 1$ is the parameter indicating the perturbation size capable of withstanding the given node without full-functionality loss.

If perturbation exceeding value C_i in one or several nodes occurs in the network, efficiency of the routes running through them changes, which leads to formation of new minimum routes. New efficiency of the graph's oriented edges is determined as:

$$\varepsilon_{ij}\left(t+1\right) = \begin{cases} \varepsilon_{ij}\left(0\right) / \frac{L_i(t)}{C_i(t)} & \text{если } L_i\left(t\right) > \mathrm{C}_i(t) \\ \varepsilon_{ij}\left(0\right) & \text{если } L_i\left(t\right) \leq \mathrm{C}_i(t) \end{cases}$$

The mean efficiency of the network is determined as [24, 25]:

$$E = \frac{1}{N_s N_n} \sum \varepsilon_{ij}$$

Here N_s is the number of source nodes, while N_n is the number of other nodes in the graph.

Then, the damage to the network can be expressed through the mean efficiency loss [26]:

$$D = \frac{E\left(\tilde{G}_0\right) - E(\tilde{G}_t)}{E\left(\tilde{G}_0\right)}$$

where $E\left(\tilde{G}_0\right)$ is the mean efficiency of the unperturbed network, $E\left(\tilde{G}_t\right)$ is the mean efficiency at time instant t.

Figure 2.5 shows the high-voltage electric network model used in this article. Yellow circles stand for sources. Green circles correspond to distribution substations. The size of a circle corresponds to the number of routes running through the node in the steady-state mode.

2.4.3 Interaction Model

Interaction of two networks has been considered through a limited number of common nodes (see Figure 2.6). As the networks are heterogeneous, the fuel and energy balance of the networks was estimated pertinent to the fuel equivalent when allowing for their interaction [27] (Table 2.3). This publication did not consider the situation when the electric network node was fully put out of action (for instance, as a result of the started fire).

Figure 2.5 High-voltage (300–400 kV) electric network model. Yellow circles – sources. Green circles – distribution substations.

Table 2.3 Factors for estimation of fuel and energy balance

Fuel	Measurement Unit	Factor of Conversion to Fuel Equivalent
Combustible natural gas	thou. cub. m.	1,154
Electric energy	thou. kW.h	0,3445

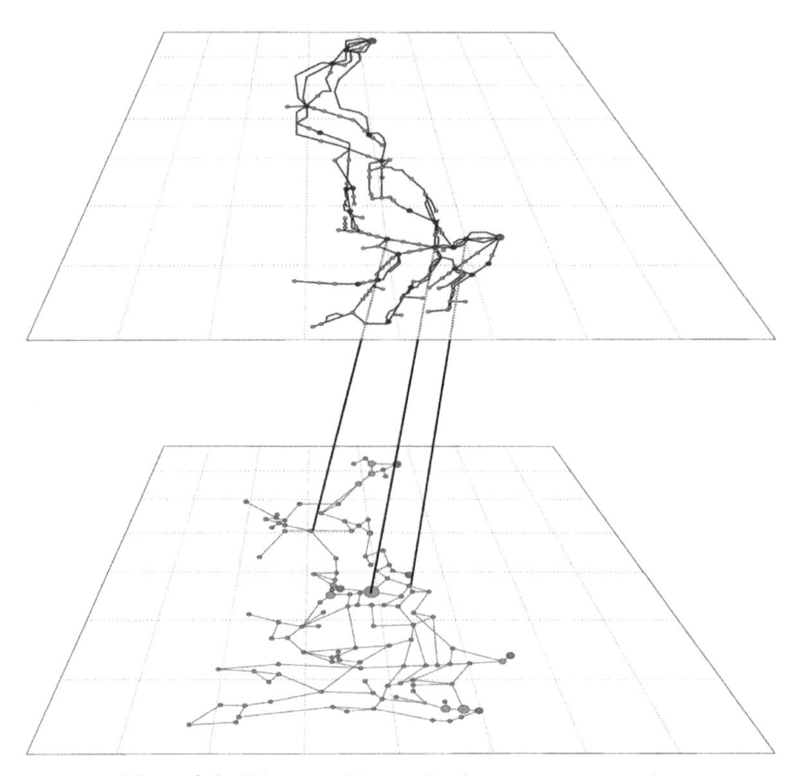

Figure 2.6 Diagram of interaction between two networks.

The responses of the electric network to gas shortfall in the common node were considered. This direction of interaction between the networks is related, above all, to the typical network perturbance times. The time of typical perturbance propagation in the gas network (the rate of gas flow in a pipe is about 10 m/s) considerably exceeds the time of perturbance propagation in the electric network: $\tau_{gas} \gg \tau_{el}$. Therefore, pertinent to the typical times of perturbance propagation in the electric network the gas transmission network can be considered to be quasistationary.

The assumption of quasistationarity of gas transmission system breaks down if we consider a possibility of destruction or full disruption of functionality of the electric network node. In this case, we should rather consider the typical times of recovery of modes of operation than typical times of perturbance propagation.

In the considered model it is expected that the gas shortfall to the common node is fully compensated by the increased electric energy consumption. This assumption in the real situation is by no means always true, but enables us to describe interaction in case of a limited number of interacting networks.

2.5 Results of Calculations

Full shut-down of the *i*-th gas-compressor station has been considered as the initial perturbance. In our publication we have used a simplified gas transmission network model. We have not considered a possibility of changing the gas production, availability of gas in the system of the underground gas storage facilities and other compensating mechanisms. In this case shut-down of the gas-compressor station leads reduction of the capacity of the respective network division and, consequently, to the necessity of redistributing the gas flows. The pressure at the outlet of the *i*–1th gas-compressor station starts rising, while the pressure at the inlet of the *i*–1th gas-compressor station starts dropping. According to Equation (2.1) this results in the pressure redistribution between the remaining nodes (see Figure 2.5) and, provided the gas transmission network integrity is retained, we have:

$$P_j < P_{crit}\ ,\ (j = 1, \ldots i - 1, i + 1, \ldots)$$

where P_{crit} is the critical pressure leading to the pipe rupture.

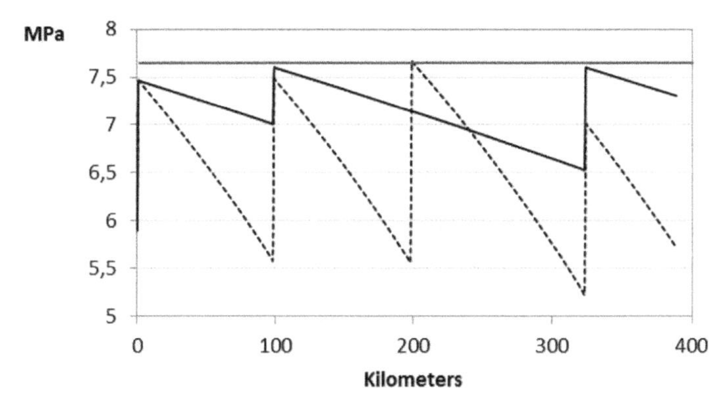

Figure 2.7 Graph of pressure variation in case of improper operation of gas-compressor station in gas transmission network *G* along one of routes from source to user. Dotted line – unperturbed mode. Solid line – mode with shut-off gas-compressor station. Red line denotes critical pressure of 7.6 MPa.

New values of the pressures in the system determine new values of the flow on the oriented edges of graph G. If obtained throughput C^*_{ij} on the cut of graph G is less than the gas consumption of this cut in the normal (failure-proof) mode, the gas shortfall takes place. Recall that the cut of the graph is essentially a set of oriented edges, whose exclusion would isolate the nodes connected by them from the network.

The response of the electric network to similar gas shortfalls to various nodes is different and depends on the network topology. Figures 2.8 and 2.9 show the dynamics of development of the perturbance that has begun in the electric network [20]. Figure 2.8 shows that the perturbance coming from the gas network has caused just a local perturbance involving a small number of the network nodes. Figure 2.9 presents the way of development of the perturbance that has already been initiated in another node and

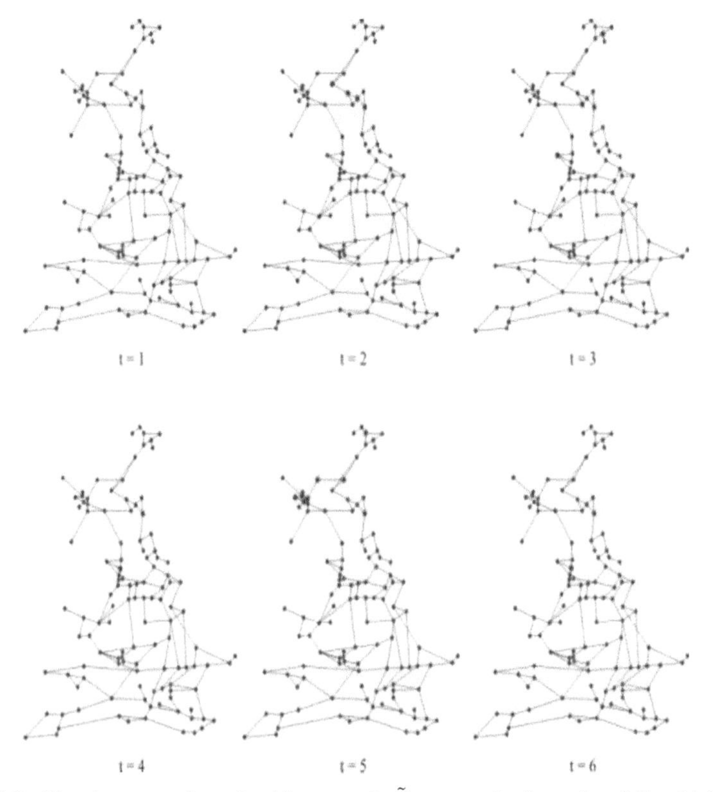

Figure 2.8 Development of overload in network $\tilde{G}$ as a result of gas shortfall, which did not lead to cascading failure. Overloaded $(L_i\,(t) > \alpha_i L_i(0))$ nodes are shown by red color [20].

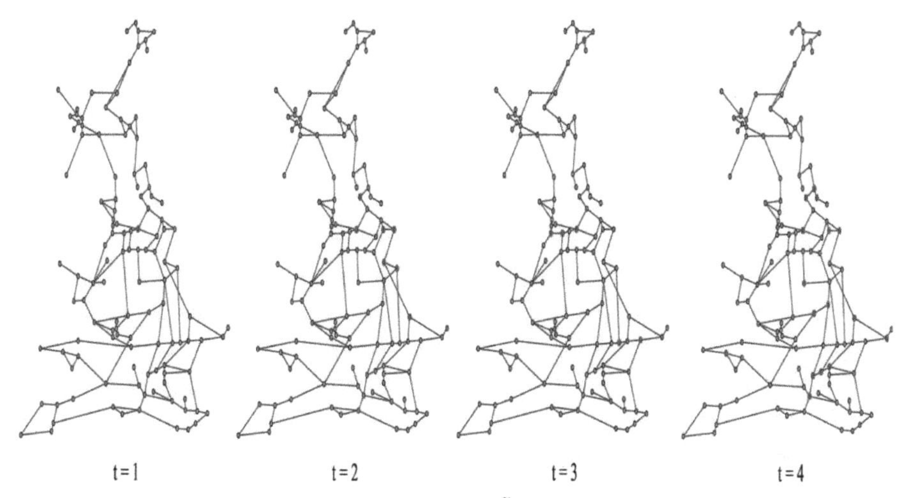

Figure 2.9 Development of overload in network $\tilde{G}$ as a result of gas shortfall, which led to cascading failure [20].

involved a substantial part of the network as a result of the faults cascade development.

The size of the perturbance of network $\tilde{G}$ depends not only from its topology, but also on its ability to withstand the overload. In the model used by us each node of network $\tilde{G}$ has common parameter α. Figure 2.10 represents the diagram of dependence of the fraction of normally functioning nodes of network $\tilde{G}$ on parameter α. As is seen, the higher the value of parameter α (i.e., the higher the ability of each node to withstand the overload), the less the damage caused by the external perturbance.

Figure 2.11 represents the graphs of damage caused to the gas (shown in the figure with a blue line) and electric (shown with a red line) networks in case of the cascading failure development. As is seen from the figure, the gas network node significance is determined not only by the gas shortfall amount, but also by the impact of this shortfall on the adjacent electric networks.

2.6 Perturbance Propagation Functions

In Subsection 2.5 some results obtained in the simple interaction model were presented. However real cascading damages can be different from these calculations. Real object of infrastructure is more robust and elastic.

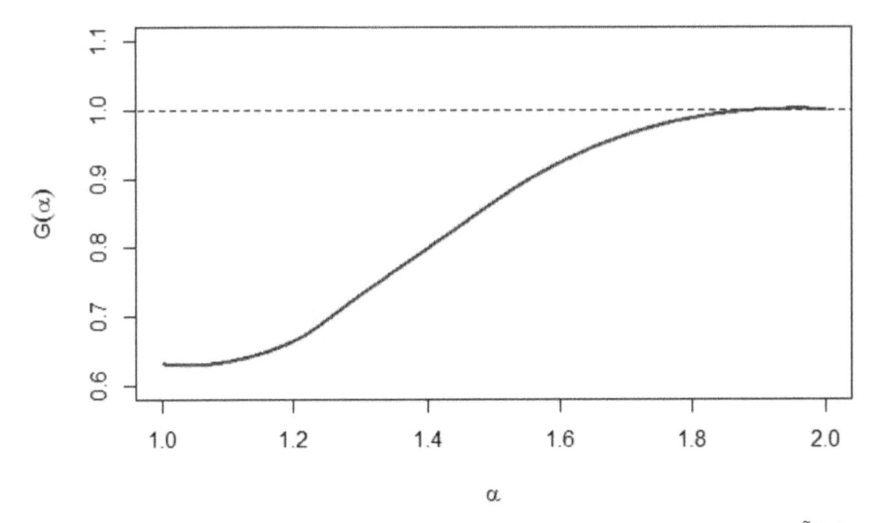

Figure 2.10 Dependence of fraction of normally functioning nodes of network $\tilde{G}(\alpha)$ on parameter indicating size of perturbance, which this node can withstand without loss of full functionality α.

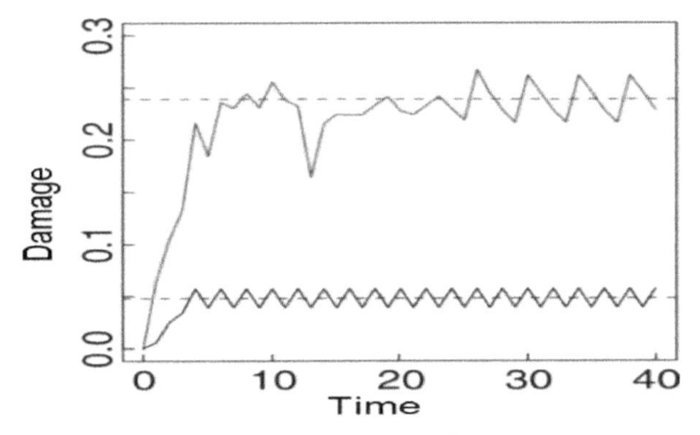

Figure 2.11 Damage in networks G (blue line) and $\tilde{G}$ (red line) during cascading failure development [20].

In real world to prevent of an accident it used the forced power reduction, the disable low priority consumers, a variety backup power sources etc. Recovery of real object is also not instantaneous.

It was proposed to describe the behavior of the real infrastructure object assigning a function of duration and amplitude of perturbance to each node (perturbance propagation function) or "translation functions". The translation

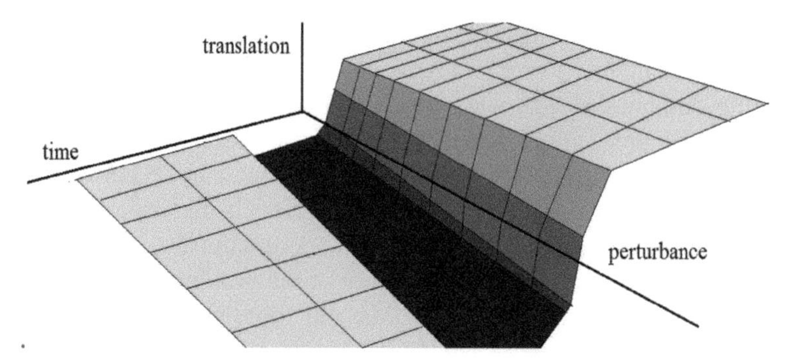

Figure 2.12 The example of the translation function type II.

functions can be very complex, and are defined by technological and siting features of objects and territories.

In the case of very accurate modeling of small networks the translation functions are tabulated for each node in network. But infrastructure networks of the infrastructurally complex territories are modeled by graphs with huge number of approximately the same nodes. In this case piecewise-linear functions are proposed for application. Let us give a brief description of these functions.

The translation function type I is simply step function. The objects which are described by this function do not work properly if capacity is not enough, but can use backup power sources.

The translation function type II is more complex (see Figure 2.12). This function can model complex reaction of object to accidents in infrastructure networks (mainly power networks). The break points and the coefficients of these functions are based on wide statistic of accidents in electric and gas transmission networks. This statistic contains data on more than 1200 accidents occurred during the period from 2011 to 2015.

And finally translation function type III is the function with the linear response to perturbance. Obviously Type I and Type III functions mathematically are the subtypes of the type II functions but described different objects in infrastructure networks.

2.7 Conclusion

The research conducted has made it possible to classify the intersystem failures occurring in critical infrastructures. The classification of ISFs by the structure of development of abnormal processes is of the greatest interest.

The intersystem failure development is simulated on the basis of the simplified model of interaction of two networks (gas supply and power supply systems). The damages to the systems caused in case of interlocking one given node in the gas supply network have been estimated. It has been demonstrated that the damages to the networks can increase if allowance is made for interaction between the networks. This effect is particularly great if the cascading failure occurs in the electric network and those adjacent to it.

With allowance for the increasing actuality of the problem of risk assessment in intersystem failures, further studies will be aimed at development of the models of the interrelated systems (transport, water supply, etc.) with due regard for the specific nature of occurrence and development of the emergency situations.

References

[1] World Urbanization Prospect (2014). *World Urbanization Prospect: The 2014 Revision, Highlights (ST/ESA/Ser.A/352)*, 28.

[2] Voropai, N. I., and Efimov, D. N. (2008). "Analysis of blackout development mechanisms in electric power systems," in *Proceedings of the IEEE Power and Energy Society General Meeting – Conversion and Delivery of Electrical Energy in the 21st Century*, 1–7.

[3] *Reduction of Risks of Cascading Failures in Electrical Power Systems*. ed. N. I. Voropai (Siberia: RAS), 302.

[4] Carreras, B. A., Lynch, V. E., Dobson, I., and Newman, D. E. (2004). Complex dynamics of blackouts in power transmission systems. Chaos 14, 643–652.

[5] Dobson, I., Carreras, B. A., and Newman, D. E. (2005). A loading-dependent model of probabilistic cascading failure. *Probab. Eng. Inf. Sci*. 19, 15–32.

[6] Dobson, I., Carreras, B. A., and Newman, D. E. (2005). "Branching process models for the exponentially increasing portions of cascading failure blackouts," in *Proceedings of the Thirty-eighth Hawaii International Conference on System Sciences*, Big Island, HI.

[7] Newman, D. E., Nkei, B., Carreras, B. A., Dobson, I., Lynch, V. E., and Gradney, P. (2005). "Risk assessment in complex interacting infrastructure systems," in *Proceedings of the Thirty-eighth Hawaii International Conference on System Sciences*, Big Island, HI.

[8] Melnikov, A. V. (2007). *Cascading Failures at Natural Gas Extraction Facilities*. LAP Lambert Academic Publishing, 224.

[9] Ganaga, S. V., and Kovalev, S. A. (2012). Simulation of failures at pipeline crossing points by means of software packages ANSYS and LS-DYNA. *Vesti Gazovoi Nauki (Gas Science News)*, 13, 133–140.
[10] Popyrin, L. S. (2000). Natural and technogenic accidents heat supply systems. *Herald Russ. Acad. Sci.* 70, 604–610.
[11] Crucitti, P., Latora, V., and Marchiori, M. (2004). Model for cascading failures in complex networks. *Phys. Rev. E.* 69, 045104.
[12] *Risk and Interdependencies in Critical Infrastructures. A Guideline for Analysis*. eds P. Hokstad, I. B. Utne, and J. Vatn (London: Springer-Verlag), 252.
[13] Pescaroli, G., and Alexander, D. (2016). Critical infrastructure, panarchies and the vulnerability paths of cascading disasters. *Nat. Disasters* 82, 175–192.
[14] Simulation von intersektoriellen Kaskadeneffekten bei Ausfällen von Versorgungsinfrastrukturen unter Verwendung des virtuellen 3D-Stadtmodells Berlins – SIMKAS-3D, Förderkennzeichen 13N10560 bis 13N10566, 2012, 87.
[15] Zio, E., and Ferrario, E. (2013). A framework for the system-of-systems analysis of the risk for a safety-critical plant exposed to external events. *Rel. Eng. Sys. Saf.* 114, 114–125.
[16] Proceedings of the 9th Future Security Research Conference, Berlin, 2014.
[17] Rinaldi, S. M., Peerenboom, J. P., and Kelly, T. K. (2001). "Identifying, understanding, and analyzing critical infrastructure interdependencies," in *Proceedings of the Control Systems, IEEE 21.6*, 11–25.
[18] Ouyang, M. (2014). Review on modeling and simulation of interdependent critical infrastructure systems. *Reliabil. Eng. Syst. Saf.* 121, 43–60.
[19] Setola, R., De Porcellinis, S., and Sforna, M. (2009). Critical infrastructure dependency assessment using the input–output inoperability model. *Int. J. Crit. Infrast. Protect.* 2.4, 170–178.
[20] Lesnykh, V. V., Petrov, V. S., and Timofeyeva, T. B. (2016). Problems of risk assessment in intersystem failures of life support facilities. *Int. J. Crit. Infrastruct.* 12, 213–228.
[21] Aliyev, R. et al. (1988). *Oil and Gas Pipeline Transport*. Moscow: Nedra Publishers.
[22] Goh, K.-I., Kahng, B., and Kim, D. (2001). *Phys. Rev. Lett.* 87:278701.
[23] Newman, M. E. J. (2001). *Phys. Rev. E* 64:016132.

[24] Asztalos, A., Sreenivasan, S., Szymanski, B. K., and Korniss, G. (2012). Distributed flow optimization and cascading effects in weighted complex networks. *Eur. Phys. J. B* 85:288.
[25] Simonsen, I., Buzna, L., Peters, K., Bornholdt, S., and Helbing, D. (2008). Dynamic effects increasing network vulnerability. *Phys. Rev. Lett.* 100, (218 701-1) - (218 701-4).
[26] Kinney, R., Crucitti, P., Albert, R., and Latora, V. (2005). Modeling cascading failures in the North American Power Grid. *Eur. Phys. J. B* 46:101.
[27] Methodological provisions for estimation of the Russian Federation fuel and energy balance in compliance with international practices. Resolution No. 46 of the RF State Committee of Statistics, 1999.

3

Stochastic Approaches to Analysis and Modeling of Multi-Sources and Big Data in Tasks of Homeland Security: Socio-Economic and Socio-Ecological Crisis Control Tools

Yuriy V. Kostyuchenko, Maxim Yuschenko and Ivan Kopachevsky

Centre for Aerospace Research of the Earth of National Academy of Sciences of Ukraine, Ukraine

Abstract

The proposed chapter is directed to the description of capabilities of applied mathematics in social and military crisis management. Crisis management of environment, infrastructure, migration, and some other issues of homeland security requires a correct base for decision making. The capabilities of stochastic methods for collecting, filtering and analyzing of multi-sources and big data in crisis situation are demonstrated in the chapter. A number of algorithms are proposed for the assessment of important social, economic, environmental issues in crisis territories. Models for social data interpretation is proposed, conclusions are discussed.

3.1 Introduction

The world around us can be described by the sets of measurable parameters. At the same time, distributions of these parameters could be described by the sets of mathematical regularities [1].

Today, there are many ways to describe and predict the behavior of observed systems and phenomena using known patterns and regularities.

In most cases, these methods are successful, and it allows to obtain reliable, accurate and timely forecasts [2].

But this way has some limitation. First of all, it is applicable to systems and processes that are ergodic and stationary [3]. That is, we indirectly assume that the statistical characteristics of the observed processes are homogeneous in time (e.g. probability density of random process is $P(x_1, x_2, \ldots, x_n, t_1, t_2, \ldots, t_n) = P(x_1, x_2, \ldots, x_n, t_1 + \Delta t, t_2 + \Delta t, \ldots, t_n + \Delta t), \forall \Delta t$, also for the expected value *E*, deviation *D*, and correlation function *R* are valid: $E(t_1) = const$, $D(t_1) = const$, $R(t_1, t_1 + \Delta t) = R(\Delta t)$, $\lim\limits_{\Delta t \to \infty} R(\Delta t) = 0$), and averaging over time is equal to averaging over the ensemble of realizations (expected value for time *T* can be assessed as: $E = \tilde{x}(t) = \frac{1}{T} \int_{-\frac{\Delta t}{2}}^{\frac{\Delta t}{2}} x(t)dt$). In this case, most mathematical models, approaches and algorithms work correctly and allow to obtain reliable results – both to study the natural and social systems.

However, in the real world the processes are not stationary and non-ergodic. And we have to either significantly change the simulation intervals, or abandon deterministic models. This is especially important while the crisis states of systems are analyzing [4].

Making of adequate and effective management decisions in crisis situations requires the availability of complete, accurate and reliable information about the state of studied system. For this, it is necessary to utilize all data from all accessible sources (Figure 3.1).

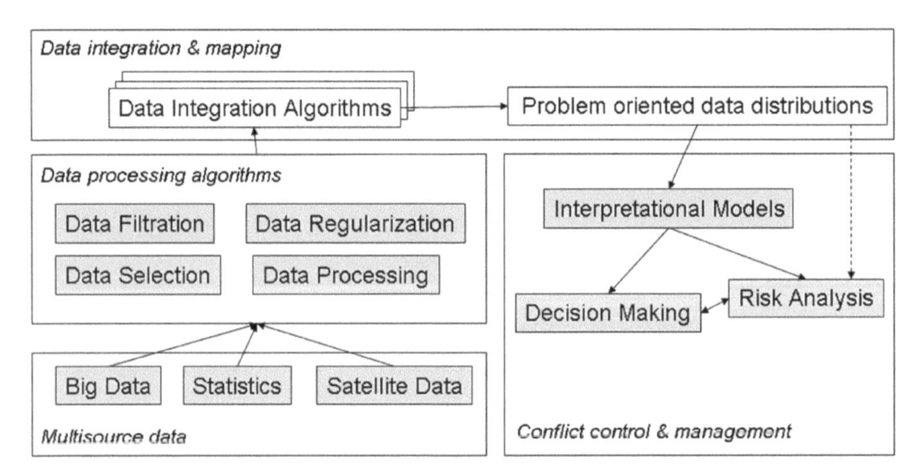

Figure 3.1 Multisource data and decision making in crisis management.

The problem is that the data from different sources have a different nature, and are characterized by different distribution properties. Thus, each type of data require a separate algorithm for processing, and their integration requires a special methodology [5].

In addition, crisis situations are characterized by significant uncertainties, both epistemic and aleatoric [6]. This significantly complicates the interpretation of results of data analysis. In this case, interpretation models could be applied. It may allow to simulate the processes occurring in the system, and to obtain the necessary forecasts of individual situations [7]. Applicability of such models is limited, but in some cases it could be useful.

A set of approaches and algorithms for the processing of data from the different sources for analysis and modeling of the social and military crisis are presented in this chapter. The task of this review is to provide a general methodology and an example of analysis individual socio-economic parameters of crisis territories.

Therefore, the proposed chapter is directed to description of capabilities of applied mathematics in social and military crisis management. Crisis management of environment, infrastructure, migration, and some other issues of homeland security requires a correct base for decision making. The capabilities of stochastic methods for collecting, filtering and analyzing of multi-sources and big data in crisis situation are demonstrated in the chapter. A number of algorithms are proposed for assessment of important social, economic, environmental issues in crisis territories. Models for social data interpretation is proposed, conclusions are discussed.

Therefore, the specific, decision making oriented way for data analysis in crisis areas is proposed, illustrated by current case study (social and military crisis in eastern Ukraine), and discussed. It is demonstrated that unusual (crisis) conditions require non-traditional approaches to data collection, analysis and interpretation. These approaches and tools, based on integrated data sources are proposed.

3.1.1 Case Study: The Conflict

The armed conflict in the Donbas (Eastern Ukraine – see Figure 3.2) extends in a spatially distributed urban agglomerations and along transport routes, e.g. with widely available and accessible infrastructure and communications. The conflict zone is an area about 16,800 square kilometers with varying number of people: from 1.9 to 2.4 million up to 3.1 in maximum, depending on the intensity of fighting, economic situation and the season. At this area in 2014

Figure 3.2 Region of conflict.

990,620 telephone lines were available, 1,786,300 Internet lines (including optical fiber), 1,196,820 Internet users, and 3,705,750 mobile phones serviced by 5 providers in 2014–2015 [8].

Region is characterized by high urbanization: more than 91,5% of population lives in cities and towns. Population of the region is aged (average age is 42,5 years, population under the age of 14 decreased by 50%, and aged 65 and over has increased by 31% during last 20 years), with short life expectance and high mortality.

The ethnic composition of the region is heterogeneous due to the migratory policy of the USSR. The ratio of ethnic Ukrainians and Russians is 55:45 in cities and 57:38 in rural areas. By language, the region is roughly homogeneous: Ukrainians and Russians are correlated as 30:70 with an increase in the share of Russians in large cities.

This region filled by archaic coal, metal and chemical industries, is a socially vulnerable area with degraded industry and infrastructure, high unemployment, high levels of inequality in income distribution and high unmotivated social benefits.

Armed conflict in the Donbas has started on April 16, 2014, when an armed group of invaders captured the city of Slovyansk and organized a tactical defense. It happened after the removal of President Yanukovych from power and Russia annexation of Crimea, and with background of widespread campaign to discredit the Ukrainian authorities in the Russian media.

Few mass protests have been organized in the region, which involved approximately 3 to 10 thousand people specially brought from the border regions of Russia. Despite the political instability the social and military situation was controlled, and formally, the social base for armed conflict was not existed.

Penetration of a group of armed militants through border to the territory of Ukraine changed the situation. Combat clashes with the police and the army began; supplying arms from Russia was started; military and political consultants arrived; and the process of organizing armed groups on the territory of Donetsk and Lugansk regions had been started.

Since May 2014 using the actual transparency of the border with the Russian Federation thousands of fighters went to Ukraine. They formed the basis of the illegal armed groups and declared the quasi-states "Lugansk People's Republic" and "Donetsk People's Republic" (L\DPR).

During 2014–2015 intensive fighting between illegal armed groups associated with L\DPR and governmental forces were detected. The most intense fighting took place in the period: May 2014–September 2014 and December 2014–February 2015.

After that, the fighting activity significantly decreased, in view of "Minsk Agreements". However, neither side considers the Agreement to be satisfactory and cannot fully perform it. Thus, despite the absence of major large-scale hostilities, the conflict continues. Currently the situation has stabilized and losses are controlled.

The main current problem is how to solve the conflict without escalation, without fighting intensifying, with losses minimization, and with improvement of security and living standards. This task requires knowing the parameters of the conflict and understanding of social dynamics of the region.

Decision-making in crises and conflicts is closely linked to the perception of threats, risks, with self-positioning, identification, social and cultural diversity. Research has shown [8, 9] that the analysis of group behavior in crisis situations should provide an analysis of many aspects: socio-economic, demographic, socio-environmental parameters, indicators of economic activity, which influence decisions.

Therefore, not only military and technical data are important for understanding of the conflict. So, methods of analysis and interpretation of interlinked military, socio-economic and socio-environmental data should be proposed and considered.

3.2 Methodological Notes: Approach to Data Analysis

In the conflicts and crises the accurate, timely and adequate information acquiring is a daunting challenge. Decision making in the conflict should be based on the all available and accessible information to be correct, adequate, useful and executable. Usually data set includes maps, GIS, satellite data, regional and local archives, complex of existing statistics, surveys and population census, etc. Also current information from local reports, media and public sources might and should be included into consideration.

Usually, parties of the conflict use different tools of censorship, all public information is distorted, and access to the crisis area for observers is usually limited. In such situation, public media and social networks became a good source of operational information in addition to the other types of data.

At the same time, using this heterogeneous composite of data sources requires a complex tool for data processing (see Figure 3.3). For example, social networks data require a correct consideration of socio-psychological and behaviorist factors. Psychological features and characteristics of group behavior of social network users have determined the distribution of data and information flows [10].

So we need to use algorithms based on comprehensive social models. Usually, when we consider the conflicts of values or cultures, and also in many cases of racial, religious and ethnic conflicts, there are stable sets of motivations of group behavior, presented in the form of symbols. They serve as a justification for violence and are widely used in propaganda. These symbols are the value-markers of certain social groups. In social networks,

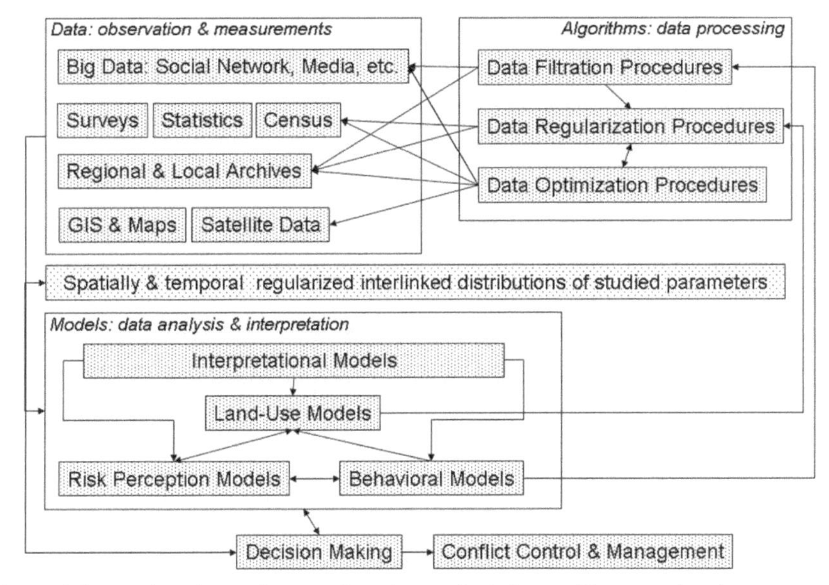

Figure 3.3 Methodology of approach to data analysis for conflict control and management.

these symbols are distributed in the form of a hashtag, and are convenient search criteria for gathering information [11].

In each case, the information contained in social networks is overabundant, with high noise, distributed by interconnected clusters. At the same time the data of statistics and surveys during conflicts are incomplete, distorted and noised.

So, we need different algorithms for different types of data (Figure 3.3), and we should apply these robust statistical algorithms for collection, filtering and classification of information using a set of interrelated criteria.

Furthermore, the classified information should be regularized/clustered to obtain spatial–temporal distributions with controlled reliability. As result of data processing, we obtain a coherent description of a situation in which all data will be statistically related to each other and mutually verified. Only in this case our data will be meaningful and have a sense as a model of conflict, rather than fragmentary information from unknown and unconfirmed sources [12].

Generally speaking, there are two types of data processing: big data filtration and multisource data optimization. The first stage is the search and collection of data by identifying indicators (for example, hashtags) on the area of conflict, location, group membership, number of killed/injured/refugees, age/date of birth, date and place of death/injuring, citizenship, etc.

3.2.1 Big Data Classification Approach

The accumulated massive statistical, survey and social network data should be classified. In other worlds, a decision making procedure should be applied to each entry of general community — the multidimensional dataset to extract a searching class ("killed non-combatants", "citizens of certain state", "killed members of illegal armed groups", etc). We propose to use for it a data classification approach based on the Bayes rule for minimum classification error in terms of maximum-a-posterior decision task in Markov random field model representation of multi-temporal, multi-source data [13]. As it was demonstrated in Ref. [8], this algorithm is working well for social network data analysis.

Let $I_j\{x_1, x_2, \ldots, x_N\}$ be given to find a record, describing a community of combatants, modeled as a set of N identically distributed n-variate random vectors of identifiers. We assume M classes $\omega_1, \omega_2, \ldots, \omega_M$ to be present among a set of combatants and we denote the resulting set of classes by $\Omega = \{\omega_1, \omega_2, \ldots, \omega_M\}$ and the class label of the k-th community member ($k = 1, 2, \ldots, N$) by $s_k \in \Omega$. By operating in the context of community classification, we assume that verified events, recognized as a training set to be available, and we denote the index set of the training community member by $T \subset \{1, 2, \ldots, N\}$ and the corresponding true class label of the k-th training community member ($k \in T$) by $s_k{*}$.

When collecting all the feature vectors of the N member of community of combatants in a single $(N \cdot n)$-dimensional column vector $X = col[x_1, x_2, \ldots, x_N]$ and all the community member labels in a discrete random vector $S = (s_1, s_2, \ldots, s_N) \in \Omega^N$, the Bayes rule approach [13] assigns to the community of combatants data X the label vector $\tilde{S}$, which maximizes the joint posterior probability $P(S|X)$:

$$\tilde{S} = \arg\max_{S \in \Omega^N} P(S|X) = \arg\max_{S \in \Omega^N} [p(X|S)P(S)] \tag{3.1}$$

where $p(X|S)$ and $P(S)$ are the joint probability density function (PDF) of the global feature vector X conditioned to the label vector S and the joint probability mass function (PMF) of the label vector itself, respectively. The Markov random field (MRF) approach offers a computationally tractable solution to this maximization problem by passing from a global model for the statistical dependence of the class labels to a model of the local general community properties, defined according to a given neighborhood system [14]. Specifically, for each k-th member of general community, a neighborhood $N_k \subset \{1, 2, \ldots, N\}$ is assumed to be defined, such that, for instance, N_k includes the four (first-order neighborhood) or the eight (second-order neighborhood) member surrounding the k-th member ($k = 1, 2, \ldots, N$). More formally, a neighborhood system is a collection $\{N_k\}_{k=1}^{N}$ of subsets of member such that each member is outside its neighborhood (i.e. $k \notin N_k \forall k = 1, 2, \ldots, N$) and neighboring member are always mutually neighbors (i.e. $k \in N_h$ and only if $h \in N_k \forall k, h = 1, 2, \ldots, N, k \neq h$). This simple and discrete topological structure attached to the general community data is exploited in the Markov random field framework to model the statistical relationships between the class labels of spatially distinct member and to provide a computationally affordable solution to the classification problem of (Equation 3.1). Specifically, we assume the feature vectors $x_1, x_2, \ldots, x_N$ to be conditionally independent and identically distributed with probability density function $p(x|s)$ ($x \in \Re^n$, $s \in \Omega$), that is [14]:

$$p(X|S) = \prod_{k=1}^{N} p(x_k|s_k) \tag{3.2}$$

and the joint prior probability mass function $P(S)$ to be a Markov random field with respect to the above-mentioned neighborhood system. The probability distribution of each k-th general community label, conditioned to all the other general community labels, is equivalent to the distribution of the k-th label conditioned only to the labels of the neighboring members ($k = 1, 2, \ldots, N$):

$$P\{s_k = \omega_i | s_h : h \neq k\} = P\{s_k = \omega_i | s_k : h \in N_k\}, i = 1, 2, \ldots, N \tag{3.3}$$

The probability mass function of S is a strictly positive function on Ω^N, so, $P(S) > 0 \forall S \in \Omega^N$.

The Markov assumption expressed by Equation (3.3) allows restricting the statistical relationships among the general community labels to the local relationships inside the predefined neighborhood, thus greatly simplifying the contextual model for the label distribution as compared to a generic global model for the joint probability mass function of all the general community labels.

Based on described approach, the Ho–Kashyap method [15] has been applied to classification of the general community.

As the result of classification we obtain a dataset with all records that meet the specified condition. For example, distribution of members of community "participant of illegal armed groups" with age, sex, social status, accessory, spatial and temporal marks inside the general community of members of illegal armed groups' combatants.

3.2.2 Multisource Data Regularization and Optimization Approach

Further, we need regularized spatial–temporal distribution of classified records (or distribution of reliable clusters). It is necessary to avoid duplication and fix falsification, inter-verify data, trace robust trends in the distributions, so get a basis for interpretation of the data. Using a two-stage procedure of data regularization may be proposed.

The method proposed is based on non-linear Kernel-based principal component algorithm (KPCA) modified according to specific data. Using this method the set of selected records has been analyzed. The robust technique of data regularization for normalization of data reliability is proposed. The technique utilizes data from different sources, different nature, and with different metrics. This approach allows to calculate regularized distributions in units invariant toward data properties and quality. Using this approach, we can analyze different types of data, regardless spatial and temporal scales and heterogeneities simultaneously.

Correct statistical analysis requires the set of data $\boldsymbol{x}_i$ with controlled reliability (above mentioned "training set"), which reflects distribution of investigated parameters during whole observation period (taking into account variances of reliability of data $\boldsymbol{x}_t$). Set of data $\boldsymbol{x}_t(\mathbf{x}_t \in R^m)$ consists of multi-source data, including data with sufficient reliability $\boldsymbol{x}_j$ $(\mathbf{x}_j \in R^m)$, where $j = 1, \ldots, N$. The problem of determination of controlled quality and reliability spatial–temporal distribution of investigated parameters $\boldsymbol{x}_i$ might be solved in framework of the tasks of multivariate random process analysis and multidimensional process regularization [16].

Required regularization may be provided by 2-stage procedure. If we are able to formulate stable hypothesis on distribution of reliability of data in the framework of defined problem we may have to propose a relatively simple way to determine investigated parameters distributions $\boldsymbol{x}_t{}^{(x,y)}$ towards distributions on measured sites $\boldsymbol{x}^m{}_t$ based on [17]:

$$x_t^{(x,y)} = \sum_{m=1}^{n} w_{x,y}(\widetilde{x}_t^m) x_t^m \tag{3.4}$$

where weighting coefficients $w_{x,y}(\widetilde{x}_t^m)$are determined as:

$$\min \left\{ \sum_{m=1}^{n} \sum_{x_t^m \in R^m} w_{x,y}(\widetilde{x}_t^m) \left(1 - \frac{x_t^m}{\widetilde{x}_t^m}\right)^2 \right\} \tag{3.5}$$

according to Ref. [17]. Here m – number of records; n – number of sources/series; $\boldsymbol{x}^m{}_t$ – distribution of data; R^m – set (aggregate collection) of data; $\widetilde{x}_t^m$ — mean distribution of searching parameters.

This is the simple way to obtain a regular spatial distribution of analyzed parameters, on which we can apply further analysis, in particular temporal regularization. At the same time, this is the first stage of regularization. This algorithm may be interpreted as the general form of Kolmogorov regularization procedure [18].

Further, should be taken into account that at second stage of regularization both data distribution temporal non-linearity (caused by imperfection of available multi-source statistics) and features of temporal–spatial heterogeneity of data distribution were caused by systemic complexity of studied community. According to Ref. [17], the Kernel-based non-linear approaches are quite effective for analysis of such types of distributions.

Proposed method is based on modified Kernel principal component analysis (KPCA). In the framework of this approach the algorithm of non-linear regularization might be described as the following rule [16]:

$$x_i = \sum_{i=1}^{N} \alpha_i^k \widetilde{k}_t(x_i, x_t) \tag{3.6}$$

In Equation (3.6) the coefficient α is selected according to optimal balance of relative validation function and covariance matrix, for example as given in Ref. [19]:

$$C^F v = \frac{1}{N} \sum_{j=1}^{N} \Phi(x_j)\Phi(x_j)^T \cdot \sum_{i=1}^{N} \alpha_i \Phi(x_i) \tag{3.7}$$

where non-linear mapping function of input data distribution Φ is determined as:

$$x_i = \sum_{i=1}^{N} \alpha_i^k \widetilde{k}_t(x_i, x_t) \tag{3.8}$$

$$\sum_{k=1}^{N} \Phi(x_k) = 0 \tag{3.9}$$

And $\widetilde{k}_t$ – is mean values of Kernel-matrix $\mathbf{K} \in R^N$ ($[\mathbf{K}]_{ij} = [k(\mathbf{x}_i, \mathbf{x}_j)]$). Vector components of matrix are determined as $\mathbf{k}_t \in R^N$; $[\mathbf{k}_i]_j = [\mathbf{k}_t(\mathbf{x}_t, \mathbf{x}_j)]$. Matrix is calculated according to modified rule: $\mathbf{k}_t(\mathbf{x}_i, \mathbf{x}_t) = \langle \rho_{j,t}^{x_j}(1-\rho_{j,i})^{x_j} \rangle$, where ρ – empirical parameters are selected according to the classification model of study community.

Using described algorithm, it is possible to obtain regularized spatial–temporal distribution of investigating parameters over the whole observation period with rectified reliability and controlled uncertainty [9, 19]. As the result we obtain a regular spatial–temporal distribution of requested parameters, prepared for the interpretation.

3.3 Population Dynamics Assessment in the Crisis Area Using Multisource Data

Correct estimation of population number and dynamics is the key issue of risk assessment and threat analysis during the conflicts. At the same time this is the complicated problem.

Panic and chaotic movement of large masses of people, paralysis of local government create deep uncertainties and make it impossible to use standard techniques and practices of population control and accounting in the conflict zone. At the same time, decision-making in situations of conflict requires reliable and operative information on the quantity, dynamics and distribution of population. To solve this problem a set of specialized algorithms and data collection tools, focused on crisis situations should be applied.

A number of stochastic algorithms, remote sensing tools, and estimation approaches might be proposed to solve this problem.

Integrated approach to population assessment, which involves number of land-use and resource management issues, is proposed in [20]. If site (x, y) is the part of mixed area with urbanized and rural districts, population in the site investigated could be presented as:

$$P_{i,t}^{(x,y)} = \frac{\mu_i^{RUR} P_i^{RUR}}{\sum_{x,y} \mu_{(x,y)}^{RUR}} + \frac{\mu_i^{UR} P_i^{UR}}{\sum_{x,y} \mu_{(x,y)}^{UR}} \tag{3.10}$$

where P_i^{RUR} – rural population, P_i^{UR} – urban population, μ^{RUR} – rural probability density coefficient, μ^{UR} – urban probability density coefficient for the certain site.

3.3.1 Population Assessment in Rural Areas

So, rural population will be determined by the rural population probability density coefficient μ_i^{RUR}, which could be defined as:

$$\mu_{i(x,y)}^{RUR} = \sum_{n,(x,y)} u_{n,(x,y)} S_{(x,y)} \tag{3.11}$$

where $u_{n,(x,y)}$ – agroecological zoning coefficient for land-use type *n* in site *(x,y)*; $S_{(x,y)}$ – square of land-use type in site *(x,y)*.

Agroecological zoning coefficient includes a number of parameters [21]:

$$u_n \to h_n(A_n; \delta)\bar{y}_n(x_n) \tag{3.12}$$

where h_n – land index, calculated for each region taking into account pollutions and soil degradation, A_n – type of land-use, δ – scaling parameter, $\bar{y}_n$ – maximum attainable yield, depends on x_n – agro-ecological condition, which includes parameters of terrain, soil, water, moisture and precipitation, climate and temperature.

Maximum attainable yield may be assessed as the functional of annual statistical yield maximum:

$$\bar{y}_n \to y \cdot (1 - u) \cdot f(k) \cdot S(T, W, R) + \Delta \tag{3.13}$$

where *u* – crop degradation index; *f(k)* – function of crop density; *S(T, W, R)* – productivity functional depends on distributions of temperature, water load and radiation; Δ – uncertainty coefficient [22].

Rural population vulnerability is determined by natural conditions, quality of lands, effectiveness of land use, intensity of pollutions, crop productivity variations during the period of crop rotation [23] and market conjuncture.

Additionally, there is a local parameter, which connects population and income distribution through variations of consumer prices of agricultural production. In the framework of general stochastic socio-economic regional model [21], a production function of “aggregate farmer” should include output index with available provincial prices p_{rc} for yield y_{rc}, the national prices p_c, including weighting coefficient w_l [24]:

$$p_r^i = w_l \frac{\sum_c p_{rc} y_{rc}}{\sum_c p_c y_{rc}} \tag{3.14}$$

where w_l determined in terms of density of urban population P_i^{UR}, coefficient of infrastructure availability (usually reflecting the road quality) β, and distance l' between the given county and all other cities and county towns as:

$$w_l = \sum_{l'} \frac{P_{l'}^{UR}}{\exp(0.01 \cdot \beta_l \cdot distance_{l,l'})} \tag{3.15}$$

This type of stochastic approach with necessary constrains and measurable variables are described and discussed in Ref. [23]. The methods to control current productivity y_{rc} as well as its variations are also proposed [25].

Therefore, a rural population density and vulnerability will depend also on distribution of urban population, in particular, on distance to city centres l, and on national distribution of crops output.

3.3.2 Population Assessment in Urban Areas

While the study area is mainly an urbanized area, we should be focused on analysis of urban population. Population on urbanized areas is distributed by other low, and it density and vulnerability should be described with other relations.

General model of urban population density p_n in region n can be presented, according to [26, 27] as:

$$p_n(r) \propto \sum_{n(x,y)} p_{n(o)} \cdot \exp\left(-\frac{r_n}{r_{n(o)}}\right)^{\sigma} \tag{3.16}$$

where $p_{n(0)}$ is the population density in the urban centre, r_n – distance of area n with localization (x,y) to centre of urbanized area, $r_{n(0)}$ – functional radius of urbanized area, σ – parameter of stage of town development.

To reduce a difference between land-use types and urban landscapes inside towns and urbanized zones, we use a fracture coefficient, according to Refs. [28, 29]:

$$\lambda_i = \sum_m \left(r_{im} - \frac{\sum_i d_{im}}{D_{im}}\right) \tag{3.17}$$

where d_{im} – size of land-use type *i* in district or town *m*, D_{im} – size of urban fracture or town *m*, included different types of land-use types d_{im}, r_{im} – distance from town *m* to the urban centre.

$$\mu_{i(x,y)}^{UR} = \sum_{n,(x,y)} \frac{r_n}{\lambda_i} A_n \ln A_{(x,y)} \exp\left(-\frac{r_n}{(A_{(x,y)}/\pi)^{1/2}}\right)^{\sigma} \quad (3.18)$$

where A_n – urbanized area, $A_{(x,y)}$ – square of town, r_n – distance to urban centre, σ – parameter of stage of town development.

Parameter of stage of urban development could be presented in a form:

$$\sigma_{n(x,y)} = \sum_{n(x,y)} \left(\frac{A_{b(x,y)} + A_{i(x,y)}^{q_m}}{A_{(x,y)}} + \beta_l \frac{l_{im}}{r_n}\right) \quad (3.19)$$

where $A_{b(x,y)}$ is a built-up area of town, $A_{i(x,y)}$ – industrial area, l_{im} – density of roads, β – coefficient of infrastructure availability (reflecting the road quality), q_m – local employment rate.

So we can conclude that vulnerability of urban population depends on distribution of urban fractures and quality urban environment: density, quality and availability of infrastructure, balance between industrial, residential and recreational zones, effectiveness of urban land use and landscape management, and social policy, particularly, employment.

3.3.3 Satellite Observations and Data Integration Approach

Final algorithm for analysis of population density should also include a component to estimate current population variations by number of statistical sources and indicators, taking into account the lack of correct statistics in the conflict zone.

In framework of this approach p_i in every spatial site (*x*,*y*) may be estimated as regular distribution:

$$p_t^{(x,y)} = \sum_{m=1}^{n} w_{x,y}(\widetilde{p}_t^m) p_t^m \quad (3.20)$$

where weighting coefficients $w_{x,y}(\widehat{p}_t^{\widetilde{m}})$ are determined as: $\min\{\sum_{m=1}^{n} \sum_{x_t^m \in R^m} w_{x,y}(\widetilde{p}_t^m)(1 - \frac{p_t^m}{\widetilde{p}_t^m})^2\}$.

Here *m* – number of records/points of measurements or observations; *n* – number of observation series; $\boldsymbol{p}^m{}_t$ – distribution of observations data;

R^m – set (aggregate collection) of observations; $\widetilde{p_t^m}$ – mean distribution of measured parameters. Set m includes m = 1, 2, 3 p_t^1– local population statistics; p_t^2 – assessment of local population $p_t^1 = \beta(P_0 - p_{refug})$ with initial regional population P_0, refugees number P_{refug} and birth rate β; p_t^3 – estimation of population number using indirect external indicators. All available sources of official and semi-official data were used [30–35].

For assessment of population by external indicators a following algorithm is proposed [36], based on satellite observations:

$$p_3 = \alpha + \beta \sum_i I_i N_i + \varepsilon_i \tag{3.21}$$

where I_i – is the radiance detected by the NASA satellite Visible Infrared Imaging Radiometer Suite (VIIRS); N_i – sum of pixels corresponds to site (*x*, *y*), α and β – empiric coefficients. Visible Infrared Imaging Radiometer Suite (VIIRS) is a 22-band radiometer to collect infrared and visible light data to observe weather, climate, oceans, wildfires, movement of ice, changes in vegetation and landforms, and monitor a nightlight. Multiyear observations of the region allow to determine coefficients α and β with enough reliability.

Based on the observed data (Figure 3.4) a form $p_3 \approx 0.03+0.47 \sum_i I_i N_i$ may be proposed.

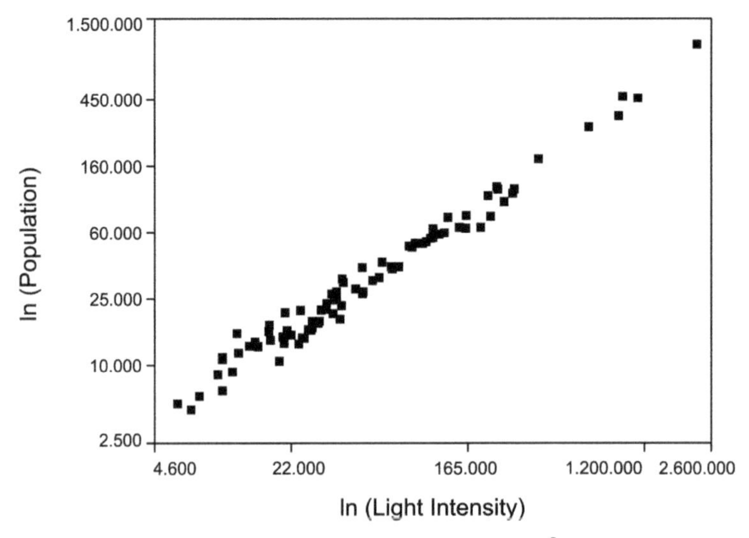

Figure 3.4 Correlation between night lights intensity (mW/cm^2/sr) and population density in the study area.

It should be noted that there is certain difference between different types of land-use in Lugansk and Donetsk regions. First of all, this area is characterized by old industrial equipments and archaic production technologies with low energy efficiency and high energy losses. Additionally, low local tariffs for electricity lead to high electricity consumption for lighting of production and coal mining sites, as well as the households. As the result, we observe the correlations of the brightness with the population at the level of 0.4–0.5, at the same time developed countries demonstrate this correlation on the level about 0.6–0.8. So about 35%–40% of observed lights – this is a loss due to the low efficiency of the regional economy. At the same time, there is a certain difference between rural and urban communities. Towns in rural area (with rural type of town planning and building density, population density, area of population employment, type of economy and land-use) demonstrate better energy efficiency. Cities in urbanized and highly industrialized areas are characterized by high energy consumption and low energy efficiency both in Donetsk and Lugansk regions with minimal distinctions. Big cities and great urban agglomeration demonstrate the same tendencies. However, it should be noted, that there are noticeable differences between Donetsk and Lugansk regions: with a similar structure of production, they are caused by different social structure and differences in traditional land use. This difference detected by satellite observations, is reflected in correlation coefficients of Equation (3.21) (see Table 3.1).

Using this algorithm (20) with conditions Equations (3.10), (3.11), (3.18), and (3.21) it is possible to calculate a distribution of population in the area studied. Result of assessment is presented in Figure 3.5.

As the data in Figure 3.5 show, the key driver of population dynamics in the conflict territory is the refugee's number. This parameter is impacted by number of economical and military factors, which should be analyzed.

Therefore, a tool for estimation of number and the control of population dynamics in crisis regions using a wide range of all available data (statistics of demography, land-use, satellite data) can be constructed.

Table 3.1 Correlation coefficients of Equation (8.21) between night lights intensity and density of population

	Donetsk Region			Lugansk Region		
	Urban	Rural	Big Cities	Urban	Rural	Big Cities
α	0,004	0,014	0,065	0,061	0,008	0,095
β	0,5	0,65	0.47	0,47	0,7	0,49
ε	0,01	0,01	0,013	0,01	0,01	0,011

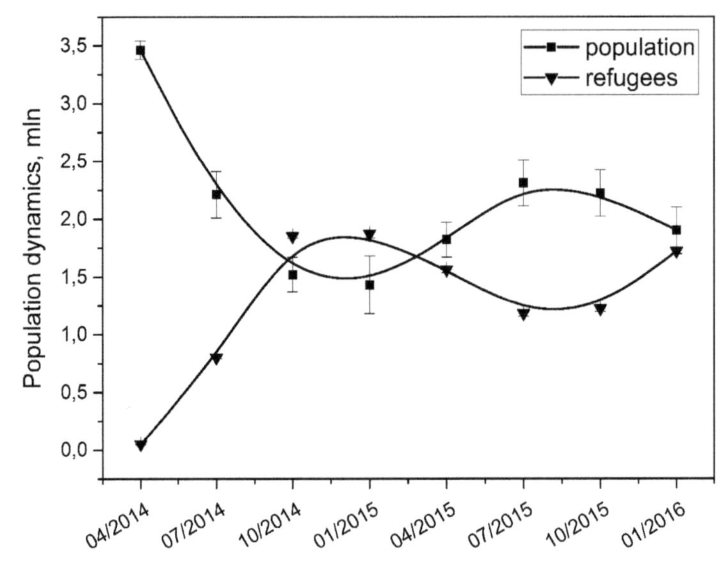

Figure 3.5 Distribution of local population and refugee's number from the crisis regions of Lugansk and Donetsk in 2014–2015.

3.4 Assessment of the Economic Dynamics in the Crisis Area Using Multisource Data

Assessment of the economic situation in the territories controlled by illegal armed groups is a challenge. Usually, there are no statistics on this area we have. Any statistics provided by the illegal administration of these areas can and should be questioned. Scraps of information provided by local residents and humanitarian organizations should be checked, regularized, filtered and verified by additional independent sources (for instance, satellite observations) for obtaining the reliable distributions.

To analyze the economy of region studied, few approaches have been used: land-use structure analysis, land cover usability and productivity analysis, and economy assessment using indirect indicators.

3.4.1 Analysis of Land-Use Structure Change: Markov's Chains Modeling of Satellite Data

Markov chains have been used to model changes in land use and land cover at a variety of spatial scales.

Markov chain models have several assumptions [37, 38]. One basic assumption is to regard land use and land cover change as a stochastic process, and different categories are the states of a chain.

A chain is defined as a stochastic process having the property that the value of the process at time t, X_t, depends only on its value at time t–1, X_{t-1}, and not on the sequence of values X_{t-2}, X_{t-3}, ..., X_0 that the process passed through in arriving at X_{t-1}. It can be expressed as:

$$P\{X_t = \alpha_j | X_0 = \alpha_0, X_1 = \alpha_1 \ldots, X_{t-1} = \alpha_i\} = P\{X_t = \alpha_j | X_{t-1} = \alpha_i\} \tag{3.22}$$

Moreover, it is convenient to regard the change process as one which is discrete in time (t = 0, 1, 2, ...).

The $P\{X_t = \alpha_j | X_{t-1} = \alpha_i\}$, known as the one-step transitional probability, gives the probability that the process makes the transition from state a_i to state a_j in one time period. When l steps are needed to implement this transition, the $P\{X_t = \alpha_j | X_{t-1} = \alpha_i\}$ is then called the l – step transition probability, $P_{ij}^{(l)}$.

If the $P_{ij}^{(l)}$ is independent of times and dependent only upon states a_i, a_j, and l, then the Markov chain is said to be homogeneous. The treatment of Markov chains in this study will be limited to first order homogeneous Markov chains. In this event:

$$P\{X_t = \alpha_j | X_{t-1} = \alpha_i\} = P_{ij} \tag{3.23}$$

where P_{ij} can be estimated from observed data by tabulating the number of times the observed data went from state i to j, n_{ij}, and by summing the number of times that state a_i occurred, n_i. Then:

$$P_{ij} = \frac{n_{ij}}{n_i} \tag{3.24}$$

As the Markov chain advances in time, the probability of being in state j after a sufficiently large number of steps becomes independent of the initial state of the chain. When this situation occurs, the chain is said to have reached a steady state. Then the limit probability, P_j, is used to determine the value of $P_{ij}^{(l)}$:

$$\lim_n P_{ij}^{(n)} = P_j \text{ where } P_j = P_i P_{ij}^{(n)}, j = 1, 2, \ldots, m, P_i = 1, P_j > 0. \tag{3.25}$$

Based on the described approach and using land use and cover change data derived from satellite images, this study establishes the validity of the Markov process for describing and analysis land use and cover changes in the study region, by examining statistical independence, Markovian compatibility, and stationarity of the data.

In performing land use and land cover change detection, a cross-tabulation detection method was employed. A change matrix was produced. Quantitative areal data of the overall land use and land cover changes as well as gains and losses in each category between 2013/14 and 2015/16 can be compiled.

Land use and land cover patterns for different periods were mapped using the Landsat Thematic Mapper and MODIS data, which have 30-m and 250-m ground resolutions. The land-use classification scheme includes the following categories: (1) urban or built-up land, (2) barren land, (3) crop-land, (4) horticulture farms (fruit trees), (5) livestock farms, (6) forest, and (7) water.

The change matrix gives the knowledge of the main types of changes (directions) in the study area. In order to analyze the nature, rate, and location of land use and land cover changes, a set of 'gains' and 'losses' images for each category was produced.

General statistical algorithm of analysis may be described as the following.

Lets define as N_{ik} – the number of cells with *i* category on the beginning of observation and with *k* at the end; and N_{ij} – number of transitions from *i* to *j* state, N_{jk} – number of transitions from *j* to *k* state during the observation period, N_j – number of cells with *j* category.

So the transition probabilities can be computed by geostatistical (GIS) analysis, and used by the following formula to calculate the expected numbers:

$$\hat{N}_{ik} = \sum_{j} \frac{(N_{ij})(N_{jk})}{N_j} \tag{3.26}$$

where $\hat{N}_{ik}$ – expected number of cells with *i* category on the beginning of observation and with *k* at the end under the Markov hypothesis.

Statistics in this case will be described as:

$$K^2 = \sum_{i,k} \frac{(N_{ik} - \hat{N}_{ik})^2}{\hat{N}_{ik}} \tag{3.27}$$

where K^2 is the distribution of Pearson's chi-squared test statistic (χ^2) with $(M-1)^2$ degrees of freedom. Statistics (χ^2) might be calculated from the relationship:

$$\chi_c^2 = \sum_{i,j} \frac{(O_{ij} - E_{ik})^2}{E_{ik}} \tag{3.28}$$

and O_{ij} – observed number of transitions, E_{ik} – expected number of transitions during observation period (Markovian distribution), here χ_c^2 is a chi-square distribution with $(m-p-1)^2$ degrees of freedom, *m* is a dimension of matrix, and *p* is a number of parameters estimated from data.

Using described approach, study area was analyzed and land-use structure changes were detected. Satellite observations allow to detect changes in land-use structure of determined classes as it presented in Table 3.2. Results of modeling of land-use structure dynamics using proposed approach are presented in Table 3.3.

As the results presented in Tables 3.2 and 3.3 demonstrate, during the period 2014–2015 no drastic changes in land-use structure of study area were

Table 3.2 Detected land-use structure changes in the study region by types

	Observation Period			Detected Change	
Land-Use Class	2014	2015	2016	(2014–15 Change)	(2015–16 Change)
Urban or built-up land, ha	43.600 (±4.300)	43.600 (±4.300)	43.600 (±4.300)	0	0
Barren land, ha	446.944	487.169	514.859	+40.225	+27.690
Cropland & agricultural lands (ha), incl.:	680.000 (±1.500)	641.000 (±1.500)	615.400 (±1.500)	–39.000	–25.600
arable land, ha	565.000 (±2.000)	531.000 (±2.000)	508.700 (±2.000)	–34.000	–21.300
grasslands, ha	34.000 (±1.700)	29.000 (±1.700)	25.700 (±1.700)	–5.000	–3.300
pastures, ha	81.000 (±2.000)	81.000 (±2.000)	81.000 (±2.000)	0	0
Horticulture farms, ha	14.280 (±1.200)	13.570 (±1.200)	12.770 (±1.200)	–710	–800
Livestock farms, ha	1.600 (±50)	1.530 (±50)	1.440 (±50)	–70	–90
Forest, ha	47.250 (±400)	46.800 (±400)	45.600 (±400)	–450	–1.200
Water, ha	3.415 (±500)	3.420 (±500)	3.420 (±500)	+5	0

Table 3.3 Detected probabilities of land-use structure changes in the study region by types

Observation Period:	2014–15	2015–16
Land-Use Class:	Detected Change	
urban or built-up land	0	0
barren land	+0,09	+0,06
cropland	–0,05	–0,04
horticulture farms	–0,03	–0,06
livestock farms	–0,04	–0,06
forest	–0,01	–0,02
water	0	0

detected. We observe decreasing of horticulture farms (4%–6%), livestock farms (5%–6%), croplands (4%–5%), and increasing of barren land (6%–9%). It shows that land-use structure demonstrates high stability toward short-term crisis changes. At the same time, using the method proposed, it is almost impossible to recognize the effectiveness of use and level of productivity on the certain type of land-use. This task should be considered and solved separately.

3.4.2 Satellite Data for Analysis of Land-Use Efficiency and Crop Structure Dynamics

Estimation of economical parameters requires not only the assessment of land-use structure but also the assessment of the crops.

In the framework of task of assessment of effectiveness of land-use and level of crop productivity, the cereal crops were estimated. For this, the methods of spectral reflectance analysis were used, based on vegetation NDVI and water NDWI spectral indices comparing. The indices were analyzed for two periods of vegetation: start of vegetation, and harvesting.

For the calculation of vegetation and water indices the following algorithms were used:

$$NDVI = \frac{NIR - RED}{NIR + RED} \quad (3.29)$$

$$NDWI = \frac{GREEN - SWIR}{GREEN + SWIR} \quad (3.30)$$

This formulae for Landsat satellite sensors has a form:

$$NDVI = \frac{ch4 - ch3}{ch4 + ch3}, NDWI = \frac{ch2 - ch7}{ch2 + ch7}, \text{ for Landsat-5,} \quad (3.31)$$

$$\text{and } NDVI = \frac{ch5 - ch4}{ch5 + ch4}, NDWI = \frac{ch3 - ch7}{ch3 + ch7} \text{ for Landsat-8.} \tag{3.32}$$

Here *ch* denotes the channel of the sensor.

To calculate changes in crop cultivation, the coefficient of spectral intensity could be proposed:

$$K_{SI} = \frac{(\sum_{x,y} NDVI_{x,y} + NDWI_{x,y})_i^{August} - (\sum_{x,y} NDVI_{x,y} + NDWI_{x,y})_i^{March\text{–}April}}{(\sum_{x,y} NDVI_{x,y} + NDWI_{x,y})_{i+1}^{August} - (\sum_{x,y} NDVI_{x,y} + NDWI_{x,y})_{i+1}^{March\text{–}April}} \tag{3.33}$$

where *i* is the number of observation season (year), *x, y* – the spatial coordinates of the fields, the sum of spectral indices is calculating in the beginning and at the end of vegetation cycle of agricultural crops – in March–April and at August.

According to algorithm (33) the changes can be calculated. Results of calculations are presented in Table 3.4.

As the results (see Table 3.4) demonstrate, the cereal crop cultivation is drastically decreased: during the conflict period usage of fields was decreasing by 83%–85%.

These results correlated well with the data of other researchers [39].

3.4.3 Data Integration Algorithm and Satellite Based Approach to Economic Activity Variations

In a conflict situation, traditional statistical information is substantially incomplete and has large uncertainties both on regional and central scales [40, 41]. However, this information is collected and processed using a correct methodology, on the basis of the available observation network, and therefore it can and should be taken into account in the analysis of the situation. At the same time, considering the substantial incompleteness of the official data, optimization and regularization algorithms should be applied to this statistics.

This data set $f_i(x)$ should be interpreted in the complex with other data sets, obtained from other sources, such as agricultural productivity, for example.

Table 3.4 Detected utilization of agricultural fields for cereal crops production in the study region

Cereal Crops Production (Relative)	2011	2012	2013	2014	2015
March–April	1	0,97	1,1	1	0,45
August	1	0,93	1,28	0,31	0,17

Other important and reliable indicator of economic activity variation is energy consumption, which could be assessed using night lights intensity assessment [42].

Using the approach [43] and data of regional observation from NASA VIIRS sensor, the following relation for GDP assessment could be proposed:

$$GDP = \sum_{i} (I_i \cdot \exp(0,46 - 0,82 \cdot \ln(N_i)) + \varepsilon_i) \tag{3.34}$$

where I_i is the detected light intensity, N_i is the square of site, and ε_i is the uncertainty coefficient.

It should be noted separately that the obtained correlations with satellite data show the low energy efficiency of regional industry. Compared with similar industrial enterprises in other countries studied by other authors [42, 43], the energy consumption for the same GDP is more by 35%–45%. This is not connected with the ongoing armed conflict, but is an indicator of archaic technologies and non-optimal production links.

Using the algorithm (34), variations of the local GDP can be estimated, and so next data set $f_i(x)$ may be obtained.

Obtained data have been optimized using regularization method based on statistical correlations within the data sets. Mapped distributions are determined as the functions $f{*}(x)$ statistically related with basic (measured, observed, estimated, etc.) points $f_i(i = 1, 2, 3, \ldots)$ $f_1(x_1), f_2(x_2), f_3(x_3), \ldots, f_n(x_n)$, n – number of assessments. This function is related with assessments through the interpolation coefficients $\lambda_1(x), \lambda_2(x), \ldots, \lambda_n(x)$ as the linear correlation:

$$f * (x) = \sum_{i}^{n} \lambda_i(x) f_i \tag{3.35}$$

Task of optimization of $f * (x)$ is the task of selection of $\lambda_1(x), \lambda_2(x), \ldots,$ $\lambda_n(x)$ coefficients. This task is solved through the search of:

$$\begin{aligned} \min \Bigg\{ & \sigma_f^2 + \bar{f}^2 - \sum_{i=1}^{n} \sum_{k=1}^{n} \lambda_i^*(x)\lambda_k^*(x)B(x_k - x_i) + \\ & + \sum_{i=1}^{n} \sum_{k=1}^{n} (\lambda_i - \lambda_i^*(x))(\lambda_k - \lambda_k^*(x))B(x_k - x_i) \Bigg\} \end{aligned} \tag{3.36}$$

where σ_{f}^2 – variance, $\bar{f}^2$ – expectation of evaluated distribution, $B(x_k - x_i)$ – autocorrelation function. This approach is resulting to rectified spatially regularized distribution of measurements x.

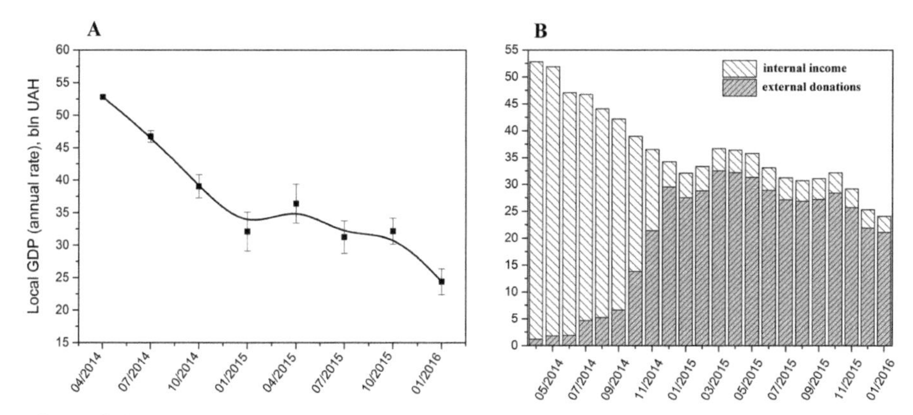

Figure 3.6 Distribution of local GDP on crisis regions of Lugansk and Donetsk in 2014–2015: (a) GDP dynamics, (b) distribution of internal and external components of GDP.

Using this algorithm with conditions, Equations (3.21), (3.22), (3.33), (3.34) it is possible to calculate a distribution of GDP in the area studied. Result of assessment is presented in Figure 3.6.

The data show that the economy of crisis territory was collapsed during first 7–8 month after the war began (Figure 3.6a). The collapse was reasoned by the fact that the local economy cannot exist outside the interlinked national economy. An analysis of local sources shows that the main tranches to the local budget come from external sources, namely, from Russia (Figure 3.6b).

Data presented in Figure 3.6 testify that in such conditions, the region cannot sustainably exist and ensure the standards of social and food security of its population.

3.5 Assessment of Number and Dynamics of Illegal Armed Groups Using Big Data

The quantitative and qualitative composition of illegal armed groups and their dynamics are important indicators of armed conflict. For the assessment of these indicators, the big data analyses approach is adopted; Equations (3.1)–(3.3) have been utilized with the regularization algorithm (4)–(9).

The described algorithms can be recognized as approaches to classification and regularization/clustering of network information. To verify information on the distribution of different social groups among investigated community, a verification approach may be applied [44, 45].

There are few parameters to evaluate clustering. Cohesiveness of clusters is interesting to evaluate clustering. In cohesive clusters, instances inside the clusters are close to each other. From the viewpoint of statistics this is equivalent to small standard deviation. In clustering, this translates to being close to the centroid of the cluster. So cohesiveness is defined as the distance from instances to the centroid of their respective clusters:

$$d_c = \sum_{i=1}^{k} \sum_{j=1}^{n(i)} \|x_j^i - c_i\|^2, \tag{3.37}$$

which is the squared distance error. Here ${x^i}_j$ is the jth instance of cluster i, $n(i)$ is the number of instances in cluster i, and c_i is the centroid of cluster i. Small values of cohesiveness d_c denote highly cohesive clusters in which all instances are close to the centroid of the cluster.

Another important measure of clustering is the distance between clusters. Separateness could also be measured by standard deviation, which is maximized when instances are far from the mean. This is equivalent to cluster centroids being far from the mean of the entire dataset:

$$d_S = \sum_{i=1}^{k} \|c - c_i\|^2, \tag{3.38}$$

where $c = \frac{1}{n} \sum_{i=1}^{n} x_i$ is the centroid of all instances and c_i is the centroid of cluster i. Large values of d_s denote clusters that are far apart.

Therefore, clustering assessment tool from Equations (3.37)–(3.38) was added to classification and regularization algorithms (3.1)–(3.9) for assessment of illegal armed groups number and dynamics in the conflict zone.

The framework of described algorithm, by the requested hashtag over 645,500 groups and individual profiles in social networks (VK, Facebook, Instagram, Twitter) in five languages (Russian, Ukrainian, English, Serbian and German) has been analyzed. Following hashtags were used: #донбасс (Donbas), #новороссия (NovoRossia), #героиновороссии (Novorossia-Heroes), #память (Memory). More than 2,780,000 entries had satisfied to determined criteria; after filtering about 248,600 entries has been collected.

Using proposed approach, required distributions of illegal armed groups' variations in the conflict territory in 2014–2015 have been calculated with grid 10 × 10 km with temporal resolution of 1 week, and reliability from 5% to 15% that depends on assessed parameter (Figures 3.7 and 3.8).

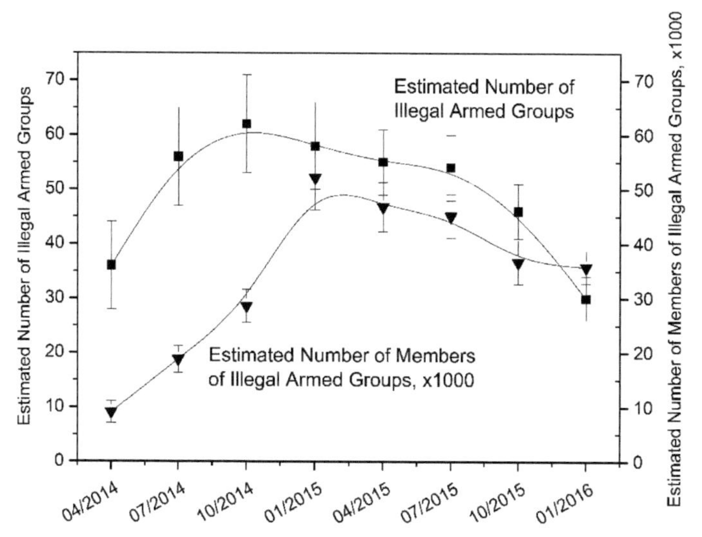

Figure 3.7 Distribution of estimated number of illegal armed groups in conflict zone in 2014–2015, and estimated number of its member's dynamics.

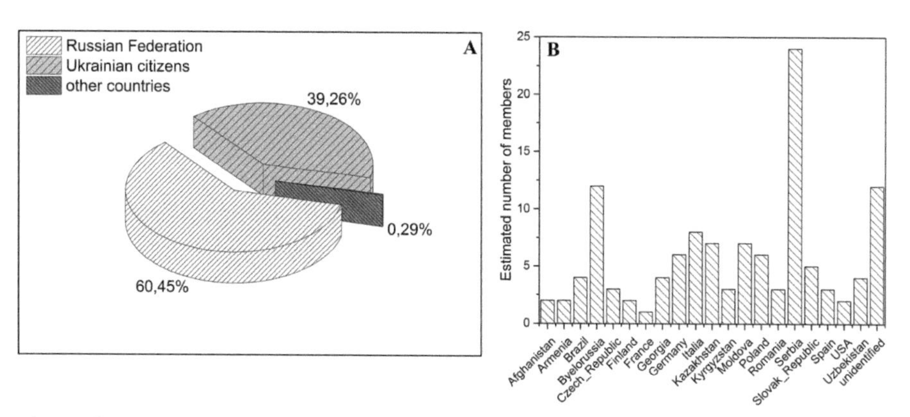

Figure 3.8 Distribution of members of illegal armed groups by citizenship: (a) general distribution, (b) third countries' citizens.

Number of illegal armed groups and number of its members (Figure 3.7) were estimated with average reliability 78%. The distribution of citizenship (Figure 3.8a,b) was detected with average reliability about 68%.

On average, in the period 2014–2015 in the study area recorded from 28 to 62, illegal armed groups were included between 5,5 and 46 thousands members (Figure 3.7). Among this community 25,100 citizens of Russian

Federations were detected including 6,200 acting personnel of Russian military forces (because Russian military troops operate in Ukraine with no badges, chevrons and without legal status, it should be recognized as a sort of illegal armed group). Besides, 16,300 Ukrainian citizens were detected and 120 citizens of other countries were recorded (Figure 3.8a).

Among third country' citizens, citizens of Afghanistan, Armenia, Brazil, Byelorussia, Czech Republic, Finland, France, Georgia (Abkhazia & Southern Ossetia), Germany, Italia, Kazakhstan, Kyrgyzstan, Moldova (Transnistria), Poland, Romania, Serbia, Slovak Republic, Spain, and USA. Besides, 6 persons with unreliable identified citizenships were detected (Figure 3.8b). It should be noted that number of third countries' combatants were changed during the study period.

This distribution demonstrates that main resources of illegal armed groups come from Russia, and so the main source of armed conflict is external.

3.6 Assessment of Combatant and Non-Combatant Losses Using Multisource Data

The problem of the loss assessment in multilateral asymmetric hybrid conflict is that the information from different network communities should be integrated with statistics from different sources. So solving the task of network community detection is required.

Considering community characteristics for community detection, the main focus is the communities that have certain group properties. Community detection is the task that is similar to the problem of clustering in data mining and machine learning. The proposed clustering techniques have proved to be useful in identifying communities in social networks. To detect separate groups, it is necessary to divide network structure into several partitions and assume these partitions represent communities. Dividing the network structure is based on the formal task of finding of an objective function, minimizing (or maximizing) that during the dividing procedure, results in a more balanced and natural partitioning of the data. This task was solved using approach, proposed in Refs. [44, 45] for hierarchical evolving communities detection.

The correctness of decision of the task of community detection is evaluating by the similar way of evaluating clustering methods in data mining. An enough precise measure to assess the correctness of community detection is the normalized mutual information (NMI) measure, which originates in information theory. Mutual information (MI) describes the amount of information that two random variables share. Therefore, by knowing one of the variables, MI measures the amount of uncertainty reduced regarding the other variable. In the case of two independent variables, the mutual information is zero. Mutual information of two variables X and Y is denoted as $I(X;Y)$. The mutual information can be used to measure the information one clustering carries regarding the available verification data. If L and H are labels and found communities; n_h and n_l are the number of data points in community h and with label l, respectively; $n_{h,l}$ is the number of nodes in community h and with label l; and n is the number of nodes, it is possible to calculate MI as:

$$MI = I(X,Y) = \sum_{h \in H} \sum_{l \in L} \frac{n_{h,l}}{n} \log \frac{n \cdot n_{h,l}}{n_h \cdot n_l} \tag{3.39}$$

MI should be normalized by mutual information measure as:

$$NMI = \frac{\sum_{h \in H} \sum_{l \in L} n_{h,l} \cdot \log \frac{n \cdot n_{h,l}}{n_h \cdot n_l}}{\sqrt{(\sum_{h \in H} n_h \log \frac{n_h}{n})(\sum_{l \in L} n_l \log \frac{n_l}{n})}} \tag{3.40}$$

An NMI value close to one indicates high similarity between communities found and labels. A value close to zero indicates a long distance between them.

After applying this algorithm, available statistics could be integrated with data from different supporting network groups: Ukrainian national forces, Russian illegal militants, and Russian-led quasi-autonomous illegal armed groups.

The statistics from official Ukrainian, Russian, and self-proclaimed L\DNR sources were used for integration procedure.

After filtration procedure more than 580,400 entries were collected from 6,220,000 entries which are satisfied to determine criteria among the 1,880,300 profiles and groups analyzed.

Using proposed approach the required distributions of losses of combatants and non-combatants in the conflict territory in 2014–2015 have been calculated with grid 10 × 10 km with temporal resolution 1 week, and reliability from 5% to 15% depends on assessed parameter.

The results of calculations are presented on Figures 3.9 and 3.10.

The losses of illegal armed groups were estimated with average reliability 76%, losses of governmental forces is estimated with average reliability 88%, and losses of non-combatants is estimated with average reliability 85% (see Figure 3.9). The distribution of citizenship of killed members of illegal armed groups (Figure 3.10) was estimated with average reliability about 72%.

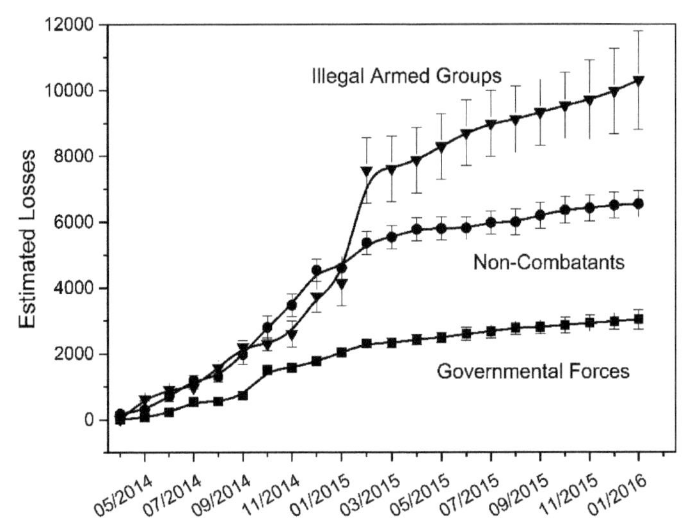

Figure 3.9 Assessment of the dynamics of losses of non-combatants, governmental forces and illegal armed groups in the period 2014–2015, using the multisource data, official statistics and network data.

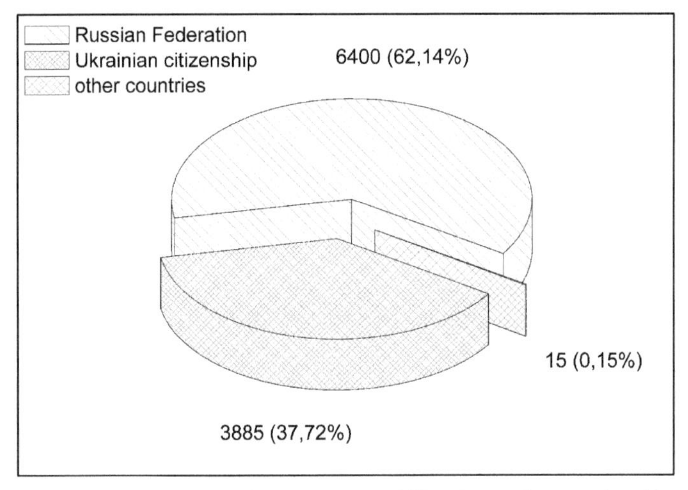

Figure 3.10 Distribution of killed members of illegal armed groups by citizenship.

Total losses of non-combatants in 2014–2015 may be estimated as 6,545 persons. Losses of governmental forces are 3,030. At the same time the losses of illegal armed groups in 2014–2015 may be assessed as 10,300 persons. Among them 6,400 citizens of Russian Federation (including 1,120 acting personnel of Russian military forces), 3,885 Ukrainian citizens, and 15 citizens of third countries (including 4 Serbians, 2 Byelorussians, and 9 persons with not reliable identified citizenship) were detected.

This distribution (Figure 3.10) generally corresponds to distribution in Figure 3.8a, and demonstrates which groups are more involved in hostilities.

At the same time the distributions in Figure 3.9 show dramatic increase of violence towards the non-combatants. This is an additional important driver of population dynamics in the crisis territories.

All detected drivers could be integrated into the interpretational model which is required for decision making.

3.7 On the Model of Population Dynamics under the Conflict

To analyze the data collected, an ecology-originated model of adaptive dynamics for structured populations can be proposed, which is aimed to study the consequences of repeated establishment of rare changes in environments set by large equilibrium population of residents, also connected with the effect of external invaders. Using adaptive dynamics instruments, it is possible to analyze how external invaders effect to residents, how residents would change their behavior (mutate) to extremists supporters, how possible resident's mutations can potentially effect to community, how mutants can invade, which successful invasions lead to the demise of original residents, and what we can do to protect original communities.

In general case, we assume that the initial resident population is stable, well-mixed, has a large size and a stable communicative links. At the same time the extremist's active supporters (mutant's) population starts up from a single manifestation (mutation), connected with external invaders. The initial magnitude of resident population makes its dynamics essentially deterministic. But invasion of external agents and invading mutants induces a strong stochastic effect. Also we assume that mutational steps are small, e.g. to allow sensible interferences, mutants should be similar to the residents. The external invaders also should be similar to the mutants.

The main tool of adaptive dynamics is invasion fitness function – the long-term average growth rate of the invaders in the community of the residents [46]. An invader with negative fitness does not gain a foothold in the given community, but a positive fitness gives a positive probability of establishment. It is important to note that when a mutant has positive invasion fitness, but because of stochasticity, its attempts for establishment fail, but this is not the end of the process. Social evolution could hide them, and later occurrences of similar mutations may become established with better luck. Theoretically, in biology and ecology, the probability of success of individual invasion attempts affects only the speed of evolution rather than its endpoint. But, in social sciences, in polymorphic diverse population or higher dimensional strategy space, the speed, direction, and outcome can be equally affected. This is the reason why analysis of invasion and mutation connected with it is so important during the crisis.

In this context the question of strategy is important. The strategy of individual denotes in this model its particular set of values for the parameters that are under social and behavioral control. These values could be presented as vector or even as the single scalar value. A polymorphic diverse community consists of individuals divided to social groups with various strategies, and we are able to consider only cases in which the number of strategies is finite.

Let's define a resident's strategy as **X** and an invader's as **Y**. Community with *N* different strategies consist of the groups 1, 2, ..., *N*, distinguished by their strategies $X_1, X_2, \ldots, X_N$. And the community of residents as a whole is indicated by $\mathbf{X} = \{X_1, X_2, \ldots, X_N\}$.

The invasion fitness function is referred as ***s***-function. A monomorphic ***s***-function $s_X(Y)$ describes the invasion fitness of a mutant with type ***Y*** in an environment set by a population of ***X***-type residents. In the case we have a community consist of *N* groups, we speak of a polymorphic ***s***-function $\boldsymbol{s}_{X1,X2,\ldots,Xn}(Y) = s_X(Y)$ to show ***N*** strategies are present.

Additionally we can define an invasion gradient: the derivative of ***s*** at a given strategy ***X*** in the mutant direction as $\frac{\partial s_X(Y)}{\partial |Y_{Y-X_1}}$.

The basic assumption of rare mutations implies in between two mutation events; the population dynamics settles at its attractor, which is unique and fixed-point. Therefore, in ***N*** group community at equilibrium in which there is a mutation in the *i*-th group, a mutant $Y \approx X_i$ is introduced. During some period the community reaches an equilibrium state again. The *N* strategies that create this new attractor depend on whether the mutant has disappeared or driven its ancestor to extinction.

Also it should be noted that there are such strategies for which the invasion gradient is zero. These strategies called "evolutionary singular strategies" – possible endpoints of evolution, non-invadable, or branching points, sources of diversity.

Traditional population models, such as Lotka–Volterra describe the communities consist of equal members [46]. In real life we should assume that communities have structure and its members are not equal. Social status and quality of life of different social groups are different. So population lives on several patches in which the resource availability and accessibility differ. Also we should assume that all individuals are independent, environment and individuals are separated, and individual behavior is a Markov process. This assumption called "Conditional linearity requirement". These assumptions are the reasons to use matrix stage models as a sub-class for Lotka–Volterra population dynamics models in our case.

To estimate the chance of mutation combined with the competition strategy we use a modified Lotka–Volterra equation:

$$\frac{d}{dt}X_i = \frac{\beta}{T} \cdot \frac{\hat{n} \cdot \mu_i(X_i)}{\sum_j u_j Var[\sum_l v_l \zeta_{lj}]} \cdot \mathbf{M}(X_i) \frac{\partial S_X(X_i)^T}{\partial Y} \tag{3.41}$$

where $\hat{n}$ is the equilibrium density of *i*-th group; β is the time to reproduce a next generation of member of the group; *T* is the time of conflict; $\mu_i(X_i)$ is probability of reproducing of member of X_i group; u_i is the number of susceptible toward a invaders impact (for example extremist propaganda) members of community; v_i is the members of community infected by invaders impact; ζ_{lj} is components of reproducing vector (mutants); and $\mathbf{M}(X_i)$ is the mutational covariance matrix of *X*.

The part $\frac{\beta}{T} \cdot \frac{\hat{n} \cdot \mu_i(X_i)}{\sum_j u_j Var[\sum_l v_l \zeta_{lj}]}$ of equation describes a process of mutation: transformation of the local population into the active and passive collaborators (economic, social and military) of invaders—illegal armed groups.

Equilibrium density of *i*-th group depends of availability and accessibility of resources and local risks as:

$$\hat{n} = n_0 \frac{\sum_i (\rho_i J_i)}{\sum_i F(X_i)} \tag{3.42}$$

There ρ_i are regional existing resources; J_i is coefficient of resources accessibility (disparity coefficient); $F(X_i)$ is the risk perception function in form of "function of willingness to pay for risk":

$$F_t(X_i) \propto \sum_{i,t=0}^{T} [V_t(\mathbf{X}_i) - b_i R_t(\mathbf{X}_i)] \tag{3.43}$$

where $F_t(X_i)$ – is the expected return of interest, risk premium, which can be interpreted as the risk perception rate; $V(\mathbf{X})$ – return of risky values $\mathbf{X} = (x_1, x_2, \ldots x_i \ldots, x_n)$; $R(\mathbf{X})$ – risk function, *t* - time; *b* – is coefficient of sensitivity known as "expected asset returns to the excess market returns".

Solution of this equation is based on strategies directed to maximize expected return of interest (find $Arg \max_{i,t}\{V_t(X_i)\}$), avoid uncertainty, maximize (find $Arg \max_{i}\{V_t(X_i)\}$), and minimize expected losses (find $Arg \min_{i}\{X_i^t\}$).

The risk function analysing in framework of the most common and most comprehensive case where the risk can be presented as the superposition of interrelated distribution function ($f(x, y)$) and damage function ($p(x, y)$):

$$R \propto \sum_{x,y} f(x, y)p(x, y) \tag{3.44}$$

Distribution function *f (x,y)* describes an impact of expanded disaster; damage function *p(x,y)* describes distribution of damaged assets: infrastructure, people, natural features, etc. in this equation (*x,y*) play a role of spatial coordinates.

Based on the prospect theory and decision making under uncertainty on cognitive bias and handling of risk [47], can be proposed a modified form of damage function as: $p(x, y|\alpha(t))$. Modified damage function includes an awareness function $\alpha(t)$, which is the superposition of risk perception function (r_p) and function of education and long-term experience (c) as: $\alpha(t) \to (c + r_p)$ following to Ref. [48].

Education function $c(t)$ describes the trend of education and experience. Risk perception function r_p which reflects security concept of human behavior, is the basis for prediction of socio-economic and socio-ecological processes. Also there is an important positive feedback of risk perception function to distribution function. Risk perception depends essentially on recent events.

The awareness function might be presented in a generalized view as follows [17]:

$$\alpha(t) = \sum_{i} (c_i + (r_p)_i) \tag{3.45}$$

Risk perception function r_p could be presented as the following uncertainty multi-source estimation [17]:

$$(r_p)_{i,t}^{(x,y)} = \delta p(\alpha) = \left| p_1(\alpha)_{i,t}^{(x,y)} - p_2(\alpha)_{i,t}^{(x,y)} \right| \Big/ \\ \Big/ \left(\frac{\sigma_\tau \sum_{\tau,m} (\frac{p_1(\alpha)_{m,\tau\in t}^{(x,y)\in M} - p_2(\alpha)_{m,\tau\in t}^{(x,y)\in M}}{2} - \sum_\tau p_1(\alpha)_{m,\tau\in t}^{(x,y)\in M}}{CoVar_m} \right) \tag{3.46}$$

where $p_1(\alpha)$ – distribution of crisis-derived losses connected with basic education level of local people, which can be presented in following form:

$$p(\alpha)_c = a_0 + p_0(x,y)\sum_{i,t}(E_{i,t}^{(x,y)} + A_{i,t}^{(x,y)}) + \xi_{i,t}^{(x,y)} \tag{3.47}$$

Here a_0 – constant coefficient; p_0 – basic level of physical losses on the site with spatial coordinates *(x,y)*; $E_{i,t}$ – education level of people group *i* in time *t* on the site *(x,y)*; $A_{i,t}$ – age of people group *i* in time *t* on the site *(x,y)*; $\xi_{i,t}^{(x,y)}$ – uncertainty coefficient.

Strategy and density of invaders ***Y*** were also described by Lotka–Volterra population model as a system of differential equations on $\Re^n : y_i = y_i f_i(y), i = 1,2,\ldots,n$, where $f : \Re^n \rightarrow \Re^n$ is smooth. Usually used standard models with $f_i = g_i + \sum_{j=1}^{n} a_{ij} y_j$, $g_i \in \Re$, and constant real matrix $A = (a_{ij})$.

We will try to include into consideration both the competition and mutualism models of interaction between resident and invaders communities. General equation in this case should be:

$$g_i + \frac{1}{\widetilde{y}_i}\sum_j((\widetilde{y}_i - y_j) - (a_{ij}x_i)) \tag{3.48}$$

where y_0 – initial quantity of invaders; g_i is the intrinsic growth (or decay) of invaders; $\widetilde{y}$ is the maximal density of external invaders population; $A = (a_{ij})$ – is the interaction matrix; x_i – density of resident population.

Using the available data and this model, it is possible to calculate and to estimate a comparative dynamics of residents (local population) and invaders (illegal armed groups number and its member), and to forecast it. Figure 3.11 demonstrates these calculations. In Figure 3.11a the results of assessments of number of illegal armed groups compared with population dynamics were presented. The mutual behavior of these parameters corresponds to the Lotka–Volterra model. In Figure 3.11b the comparison of observation and model data of local population and invaders is presented. Figure 3.11b demonstrates a good corresponding of the proposed model to observed data.

The presented model allows to make several useful conclusions. In particular, to destabilize the region (with a high level of urbanization, with archaic industry and infrastructure) with a population about 3 million people (aged, highly unemployed and socially disoriented), it is enough to impact by several groups of invaders with total size about 0.7–1.3 thousand (up to 0.05% of the resident population).

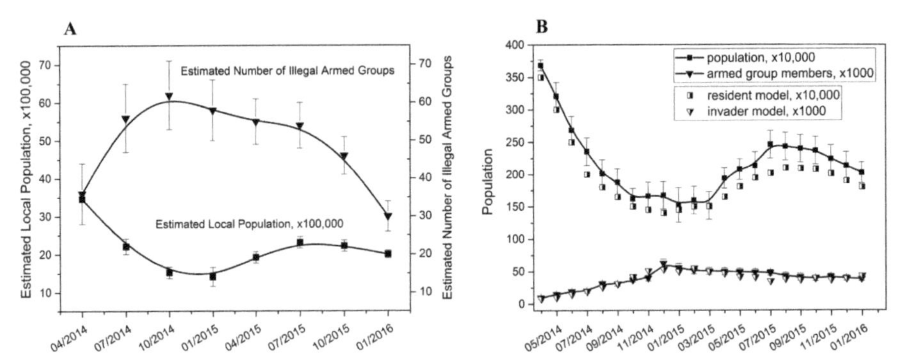

Figure 3.11 Comparison of observed data (a) and results of adaptive dynamics modeling of structured populations using a modified stochastic Lotka–Volterra approach (b).

To support the social and military crisis in the "hot" state with sufficient mobilization, it is necessary to provide 0.8–1.2% of invaders in the resident population.

The studied resident population is sensitive to changes in the number of invaders at level of 3.5%–5%. In other words, the change in the number of illegal armed groups by 1–1.5 thousand causes a detectable change in the resident population.

In the situation of modern wars, with aggressive propaganda, demonstrated and hidden links between facts (real and fictional) are playing a greater role than the data or facts as such. Therefore, the adequate interpretational models aimed to identification of correct links between parameters of the systems studied (which are detected in form of indicators) and obtaining of reliable parameters distribution are important.

Proposed form of interpretational model allows to analyze whole complex of available data and to investigate of social dynamics of the crisis region. It allows to understand how local population reacts to invaders and their varied impacts. This model could be useful for the decision making and policy planning in the conflict control.

3.8 Concluding Remarks

Changing nature of modern conflicts requires novel approaches to its analysis and decision making [49, 50]. New approaches require new data and algorithms for its processing. If tradition conflicts were driven by geo-political, economic or ideological interest, modern conflicts are the wars of identities, which have the internal, behavioristic reasons and drivers [51]. Because the rise of identity politics is associated with new communications technologies [49], the analysis of identity associated conflicts should include analysis of data from communicative networks.

So the multisource data including social network and other big data should be involved into consideration. It requires development of methodology as well as a complex of interlinked algorithms for multisource varied nature data processing and analysis. These approaches and algorithms should be focused on wide range of data, including issues of group behavior.

The proposed set of approaches and algorithms for processing of data from the different sources for analysis and modeling of the social and military crisis demonstrates the high capabilities of applied mathematics in social and military crisis management. The capabilities of stochastic methods for collecting, filtering and analyzing of multi-sources and big data in crisis situation are demonstrated. A number of algorithms for assessment of important social, economic, environmental issues in crisis territories are proposed.

Besides, decision-making in crises and conflicts should be based on analysis of the perception of threats and risks, with understanding of self-positioning and identification of people of the social and cultural population diversity. Research has shown [8] that the analysis of group behavior in crisis situations should provide an analysis of many aspects: socio-economic, demographic, socio-environmental parameters, indicators of economic activity, which influence decisions. Models for social data interpretation are proposed.

Therefore, this study shows that, for the requirement of modern conflict analysis, a set of algorithms can be developed and the set of interpretative models as well as the reliable data distributions may be obtained in high uncertainty environments.

Proposed approaches to data analysis allow to estimate a correctness and reliability of existing conflict statistics. Reports of Ukrainian governmental bodies on refugees and displaced persons number have a reliability on the level of 78%–82%, and may be corrected with the data of volunteer organization and social media reports up to the level of 92%–95%. Data on losses of non-combatants are correct with reliability of 93%–95% on the territories controlled by government. Data on the losses of combatants have the reliability of 82%–85%, and may by corrected to the level 91%–94% with data of volunteer organization and social media reports. Reports of leaders of illegal armed groups and representatives of illegal quasi-states are reliable on the level 35%–45% for all categories of data. Using proposed data regularization and model-based interpretation approach this parameter may be increased up to 76%–82%. Data from official Russian sources do not exist, so its initial reliability can't be assessed. Therefore, it is possible to increase reliability of conflict statistics by 10%–25% using proposed approach.

Obtained distributions of social, economic and military data could allow to make few important conclusions concerning the analyzed conflict.

First of all, the analyzed conflict in the Donbas cannot be described as an internal or civil conflict in Ukraine in any case. It should be analyzed as a form of aggression of Russia against Ukraine, and thus should be interpreted as a special form of asymmetric conflict [52] by using hybrid instruments [53] between Ukraine and Russia [54]. This is evidenced by the resources supply: more than 80% of the financial, and more than 60% of military, technical and human resources of illegal armed groups is supplied by Russia. The external character of the conflict is confirmed by distribution of detected losses of illegal armed groups, more than 60% of which are citizens of Russia and Russian military personnel.

It may be noted, that the conflict has not clearly expressed the ethnic or linguistic attributes. There are weak indicators of socio-cultural differences, expressed in the patterns of behavior and perception of threats, and are also detected in land-use features. The biggest indicator of conflict driver is the logistics parameters: the conflict zone extends to 35–40 km along and around the roots of delivery from the territory of Russian Federation, the resources for illegal armed groups.

As the population model shows, the propaganda should be recognized as an instrument of aggression. Propaganda is a way to recruiting local population to collaboration with hybrid occupational forces.

Analysing the statistics of conflicts, we should remember that we are dealing with human suffering, so this activity requires accuracy and responsibility. Taking into account the nature of modern conflict, the most important task of data analysis and interpretation is the comprehensive and equi-resistance protection of civil population.

The task of the conflict solving without further escalation, fighting intensifying, with losses minimization, and with improvement of security and living standards, requires the understanding of parameters of the conflict and the social dynamics of the region. The proposed methodology and the set of interlinked algorithms of data processing are directed to the solving of this task.

Acknowledgements

The authors are grateful to anonymous referees for constructive suggestions that resulted in important improvements to the chapter, to colleagues from the International Institute for Applied Systems Analysis (IIASA), from the American Statistical Association (ASA), American Meteorological Society (AMS), and from the International Association for Promoting Geoethics (IAPG) for their critical and constructive comments and suggestions. The authors express their particular thanks to Science and Technology Center in Ukraine (STCU) for partial support of this study in the framework of research project #6165 "Information and Technological Support for Greenhouse Effect Impact Assessment on Regional Climate using Remote Sensing", as well as to National Academy of Sciences of Ukraine for partial support of this study in the framework of joint Ukraine-IIASA research program "Integrated modeling of food, energy and water security management for sustainable social, economic and environmental developments".

References

[1] Siegler, R. S., and Opfer, J. E. (2003). The development of numerical estimation: evidence for multiple representations of numerical quantity. *Psychol. Sci.* 14, 237–250.

[2] Strang, G., and Aarikka, K. (1986). *Introduction to Applied Mathematics*, Vol. 16. Wellesley, MA: Wellesley-Cambridge Press.

[3] Gentle, J. E., Härdle, W. K., and Mori, Y. (eds). (2004). *Handbook of Computational Statistics: Concepts and Methods.* Berlin: Springer Science & Business Media.

[4] Lemert, C. C. (2015). *Sociology After the Crisis.* Abingdon: Routledge.

[5] Burrows, R., and Savage, M. (2014). After the crisis? Big data and the methodological challenges of empirical sociology. *Big Data Soc.* 1, 28–35.

[6] Bryman, A., and Burgess, B. (eds). (2002). *Analyzing Qualitative Data.* Abingdon: Routledge.

[7] Ciborra, C. U. (1998). Crisis and foundations: an inquiry into the nature and limits of models and methods in the information systems discipline. *J. Strateg. Inform. Syst.* 7, 5–16.

[8] Kostyuchenko, Y. V., and Yuschenko, M. (2016). "Methods and tools of big data analysis for terroristic behavior study and threat identification: illegal armed groups during the conflict in Donbass region (East

Ukraine) in period 2014–2015," in *Threat Mitigation and Detection of Cyber Warfare and Terrorism Activities*, ed. M. E. Korstanje (Hershey, PA: IGI Global), 52–66. doi: 10.4018/978-1-5225-1938-6.ch003

[9] Kostyuchenko, Y. V. (2016). "Risk perception based approach to analysis of social vulnerability," in *Risk Perception: Theories and Approaches*, ed. T. Spencer (Hauppauge, NY: Nova Science Publishers).

[10] Lerman, K. (2013). "Social informatics: using big data to understand social behavior," in *Handbook of Human Computation*, ed. P. Micheluccipp (New York, NY: Springer), 751–759.

[11] Sultana, M., Paul, P. P., and Gavrilova, M. (2015). Social behavioral biometrics: an emerging trend. *Int. J. Pattern Recogn. Artif. Intell.* 29:1556013.

[12] Caverlee, J., Cheng, Z., Sui, D. Z., and Kamath, K. Y. (2013). Towards geo-social intelligence: mining, analyzing, and leveraging geospatial footprints in social media. *IEEE Data Eng. Bull.* 36, 33–41.

[13] Duda, R. O., Hart, P. E., and Stork, D. G. (2012). *Pattern Classification*. Hoboken, NJ: John Wiley & Sons.

[14] Geman, S., and Geman, D. (1993). Stochastic relaxation, Gibbs distributions and the Bayesian restoration of images. *J. Appl. Stat.* 20, 25–62.

[15] Ho, Y. C., and Kashyap, R. L. (1965). An algorithm for linear inequalities and its applications. *IEEE Trans. Electron. Comput.* 5, 683–688.

[16] Kostyuchenko, Y., Movchan, D., Kopachevsky, I., Bilous, Y. (2015). "Robust algorithm of multi-source data analysis for evaluation of social vulnerability in risk assessment tasks," in *Proceedings of the SAI Intel System*, London.

[17] Kostyuchenko, Y., Movchan, D. (2015). "Quantitative parameter of risk perception: can we measure a geoethic and socio-economic component in disaster vulnerability?," in *Geoethics: The Role and Responsibility of Geoscientists*, eds S. Peppoloni and G. Di Capua (London: Geological Society), 4–19.

[18] Ermoliev, Y., Makowski, M., and Marti, K. (2012). *Managing Safety of Heterogeneous Systems*. Berlin: Springer.

[19] Kostyuchenko, Y. V. (2015). "Geostatistics and remote sensing for extremes forecasting and disaster risk multiscale analysis," in *Numerical Methods for Reliability and Safety Assessment: Multiscale and Multiphysics Systems*, eds S. Kadry and A. El Hami (Cham: Springer International Publishing), 404–423.

[20] Kopachevsky, I., Kostyuchenko, Y. V., and Stoyka, O. (2016). Land use drivers of population dynamics in tasks of security management and risk assessment. *Int. J. Math. Eng. Manag. Sci.* 1, 18–24.

[21] Fischer, G., Van Velthuizen, H. T., Shah, M. M., and Nachtergaele, F. O. (2002). *Global Agro-ecological Assessment for Agriculture in the 21st Century: Methodology and Results*. Laxenburg: IIASA.

[22] Gommes, R., Acunzo, M., Baas, S., Bernardi, M., Jost, S., Mukhala, E., et al. (2010). "Communication approaches in applied agrometeorology," in *Applied Agrometeorology,* ed. K. Stigter (Berlin: Springer), 263–286.

[23] Movchan, D., Kostyuchenko, Y., Marton, L., Frayer, O., and Kyryzyuk, S. (2014). "Uncertainty analysis in crop productivity and remote estimation for agricultural risk assessment," in *Vulnerability, Uncertainty, and Risk: Quantification, Mitigation, and Management*, eds M. Beer, S.-K. Au, and J. W. Hall (Liverpool: ASCE), 1008–1015.

[24] Albersen, P., Fischer, G., and Keyzer, M. L. (2002). *Sun, Estimation of Agricultural Production Relations in the LUC Model for China.* IIASA Report RR-02-03. Laxenburg: International Institute for Applied Systems Analysis.

[25] Kostyuchenko, Y., Bilous, Y., Movchan, D., Márton, L., and Kopachevsky, I. (2013) Toward methodology of satellite observation utilization for agricultural production risk assessment. *IERI Proc.* 5, 21–27.

[26] Clark, C. (1951). Urban population densities. *J. R. Stat. Soc.* 114, 490–496.

[27] Chen, Y. G., (2008). A wave-spectrum analysis of urban population density: entropy, fractal, and spatial localization. *Discrete Dyn. Nat. Soc.* 2008:728420.

[28] White, R., and Engelen, G. (1994). Urban systems dynamics and cellular automata: fractal structures between order and chaos. *Chaos Solitons Fractals* 4, 563–583.

[29] Chen, Y. G., and Zhou, Y. X. (2008). Scaling laws and indications of self-organized criticality in urban systems. *Chaos Soliton Fractals* 35, 85–98.

[30] Interdepartmental Coordination Headquarters of the State Emergency Service of Ukraine (2017). *Support of Social Security of Ukrainian Citizens Moved from the Territories of Antiterroristic Operation and from Temporally Occupied Territories*. Available at: http://cn.dsns.gov.ua/ua/Mizhvidomchiy-koordinaciyniy-shtab.html

[31] Main DataBase of Internally Displaced Persons of Ministry of Social Policy of Ukraine (2017). Available at: http://www.msp.gov.ua/en/timeline/Vnutrishno-peremishcheni-osobi.html

[32] State Committee of Statistics of Ukraine. Population Statistics of Ukraine (2017). Available at: http://database.ukrcensus.gov.ua/MULT/Dialog/statfile_c.asp

[33] Main Department of Statistics of Donetsk People's Republic (2017). Available at: http://glavstat.govdnr.ru

[34] Main Department of Statistics of Lugansk People's Republic (2017). Available at: http://www.gkslnr.su/stat_info

[35] Federal State Statistics Service of Russian Federation, Number and Migration of the Population of the Russian Federation (2017). Available at: http://www.gks.ru/wps/wcm/connect/rosstat_main/rosstat/ru/statistics/publications/catalog/doc_1140096034906

[36] Levin, N., and Duke, Y. (2012). High spatial resolution night-time light images for demographic and socio-economic studies. *Remote Sens. Environ.* 119, 1–10.

[37] Parzen, E. (1999). *Stochastic Processes*. Philadelphia, PA: Society for Industrial and Applied Mathematics.

[38] Stewart, W. J. (1994). *Introduction to the Numerical Solution of Markov Chains*. Princeton, NJ: Princeton University Press.

[39] Lyalko, V. I., Elistratova, L. O., and Apostolov, O. A. (2017). *Estimation of Areas of Winter Crops by the Data of Space Survey from the Landsat Satellite in the Donetsk Region*. Kiev: National Academy of Sciences Ukraine.

[40] Main Department of Statistics of Ukraine in Donetsk Region (2017). Available at: http://www.donetskstat.gov.ua/statinform1/ec_activity.php

[41] Main Department of Statistics of Ukraine in Lugansk Region (2017). Available at: http://www.lg.ukrstat.gov.ua/statinform.php.htm

[42] Sutton, P. C., Elvidge, C. D., and Ghosh, T. (2007). Estimation of gross domestic product at sub-national scales using nighttime satellite imagery. *Int. J. Ecol. Econ. Stat.* 8, 5–21.

[43] Doll, C. N., Muller, J. P., and Morley, J. G. (2006). Mapping regional economic activity from night-time light satellite imagery. *Ecol. Econ.* 57, 75–92.

[44] Zafarani, R., Mohammad, A., A., and Liu H. (2014). *Social Media Mining: An Introduction*. Cambridge: Cambridge University Press.

[45] Guandong, X. (2013). *Social Media Mining and Social Network Analysis: Emerging Research*. Hershey, PA: IGI Global.

[46] Haccou, P., Jagers, P., and Vatutin, V. A. (2005). *Branching Processes: Variation, growth, and Extinction of Populations*. Cambridge: Cambridge University Press.
[47] Kahneman, D., and Tversky, A. (eds). (2000). *Choices, Values, and Frames*. Cambridge: University Press.
[48] Tversky, A., and Kahneman, D. (1985). *The Framing of Decisions and the Psychology of Choice, in Environmental Impact Assessment, Technology Assessment, and Risk Analysis*. Berlin: Springer, 7–129.
[49] Kaldor, M. (2013). In defence of new wars: stability. *Int. J. Sec. Dev.* 2, 1–16.
[50] Duffield, M. (2001). *Global Governance and the New Wars: The Merging of Security and Development*. London: Zed Books.
[51] Münkler, H. (2005). *The New Wars*. Cambridge: Polity.
[52] Gross, M. L. (2009). Asymmetric war, symmetrical intentions: killing civilians in modern armed conflict *Glob. Crime* 10, 320–336.
[53] Schroefl, J., and Kaufman, S. J. (2014). Hybrid actors, tactical variety: rethinking asymmetric and hybrid war. *Stud. Conflict Terror.* 37, 862–880.
[54] Kofman, M., and Rojansky, M. (2015). *A Closer Look at Russias "Hybrid War"*. Washington, DC: Kennan Cable.

4

Modeling and Performance Evaluation of Computational DoS Attack on an Access Point in Wireless LANs

Rajeev Singh[1] **and Teek Parval Sharma**[2]

[1]Department of Computer Engineering, G.B. Pant University of Agriculture and Technology, Pantnagar, India
[2]Department of Computer Engineering, National Institute of Technology, Hamirpur (H.P.), India

Abstract

A computational DoS attack is one where attacker continuously floods a node by large number of packets requiring excessive processing. In WLANs, data frames are usually protected by cryptographic primitives like Message Integrity Code (MIC). A large number of attack frames bearing wrong MIC, make the recipient node busy in processing all the time and dropping the legitimate frames that overflow from node's buffer. Such attack is usually carried towards an Access Point (AP) and is effective at AP because the attack frames can only be distinguished from legitimate ones after processing. In this work, a scenario involving processing at AP is modeled in network simulator (NS2). Using this scenario we show that the computational flooding DoS attack is successful via data frame flooding on AP and the attack is able to disrupt the wireless station communication. Key Hiding Communication (KHC) scheme is proposed as a secure and lightweight communication for WLANs. We also validate the scheme's ability to maintain communication with the wireless station during computational DoS attack. Using the same simulation scenario we show that KHC scheme is effective against computational flooding DoS attack conducted via data frame flooding on AP and it is able to maintain the communication. Both Transmission Control Protocol (TCP)-based communication and User Datagram Protocol (UDP)-based communication are considered during evaluation.

4.1 Introduction

An Access Point (AP) is an example of critical communication system that offers services to a small area. The most important services are Internet and data transfer. Denial of Service (DoS) attack is one of the serious concerns in WLAN security. DoS attack targets availability which is concerned with delivery of required services without failure. It is crucial and of utmost importance to maintain availability in AP-based systems. DoS problem in WLANs is practical and is not a wild cry. It is increasing day by day [1–4]. Easy availability of DoS attack tools and mechanisms are mainly responsible for it [5–13]. The security protocols like Wired Equivalent Privacy (WEP), Wi-Fi Protected Access (WPA) and IEEE 802.11i [14] (WPA2) have no considerations for the DoS attacks [1–4, 15–21]. The WLAN communication process is divided into medium access, connection, entity authentication, data communication and disconnection procedures. DoS attack in medium access procedures is performed using unprotected control frames like RTS and CTS [22, 23]; DoS attack in connection and disconnection procedures is performed using unprotected management frames like authentication, deauthentication, association and disassociation [24–29]; DoS attack in entity authentication and in data communication procedures is performed using data frames [30, 31].

Among several kinds of DoS attacks, flooding DoS attacks are most common. In the initial communication setup phases it can be conducted by using frames like RTS, authentication, association, etc. Recently, flooding DoS attack in data communication procedures is also reported using protected data frames [32]. In this attack, attacker floods the node by large number of data frames continuously. Current research shows that this attack is possible at an Access Point due to substantial and continuous occupancy of Medium Access Control (MAC) buffer by the attack frames. We aim that along with MAC buffer limitation, the processing required per frame at AP for protective measures like authentication, integrity, etc. is also responsible for it. This attack where attacker utilizes the computing resource is termed as computational DoS attack. During data communication, per frame authentication is proposed as the solution to flooding DoS attack. An Access Point in Wireless LAN (WLAN) has limited capacity and limited resources, e.g. processing power and memory. Hence, AP can fell an easy prey to DoS attack as its buffer can be easily flooded and blocked by the attack frames. In the current WLAN security protocols like IEEE 802.11i [14], data frame authentication is provided by Message Integrity Code (MIC) calculation and verification [33-35]. MIC calculation and verification require processing time at the AP. If attacker is able to send large number of frames for MIC verification, then this may lead to computational DoS attack [32]. In this attack AP is busy in MIC verification and hence frames in the MAC buffer increase, leading to overflow, frame rejection and retransmission. Obviously MIC verification of attack frames will fail as the only intension of attacker is to make AP busy all the time during attack leading to dropping of regular data frames. We show that even small processing delay is harmful and may lead to computational DoS attack.

Most of the simulation-based studies for flooding DoS attack in WLANs consider the buffer overflow but neglect the processing involved at AP [22, 23, 36, 37]. The NS2 simulator-based [38–40] case studies model flooding DoS attack based upon the queue overflow while neglecting the processing requirements [37]. NS2 procedures consider buffering of a frame at AP due to frame transmission delay and not due to processing delay. Thus in existing NS2 models, the regular and attack packets queued up due to transmission delay. We add processing delay per frame in this model for making it more realistic and practical. This addition of processing at the AP will also help in near future to improve the simulated cryptographic studies targeting throughput and delay in WLANs.

Key Hiding Communication (KHC) scheme is proposed in Ref. [32], as a secure and lightweight communication for WLANs. In this paper we also validate the scheme's ability to maintain communication with wireless station during the attack. In this scheme for verifying the frame authenticity, first fresh key and initial vector (IV) are evolved. Then they are compared with the key and IV obtained from received communication frame. The fresh key evolving per frame take time of the order of 0.02 ms. After successful comparison, the frame integrity is verified. The major reason that attributed to success of the KHC against computational DoS attack is the ability to authenticate the data frames in lightweight manner and to verify the integrity of only those frames that are able to pass the authentication check.

Contributions of the paper are: (1) development of realistic WLAN simulation model with processing delay at AP in NS2, (2) realization of the fact that in flooding DoS attack using data frames, the processing delay at AP worsens the situation and, (3) validation of the KHC scheme against the computational DoS attack. Proposed work helps in analysing the effect of varying processing time and observing that computational DoS attack is effective in restricting the ongoing wireless communication at AP in absence of any protection method. Proposed work establishes usefulness of the KHC scheme against computational DoS attack, i.e. the ongoing wireless communication remains unaffected under the attack. In this paper we vary the processing delay at AP as 0.2 and 0.02 ms. Two different types of user traffic, i.e. File Transfer Protocol (FTP) and Constant bit Rate (CBR) are considered. FTP utilizes Transmission Control Protocol (TCP) for reliable communication. TCP is a reactive protocol and tries to control the sender's flow rate depending upon packet drops and congestion in the path. CBR utilizes User Datagram Protocol (UDP). We consider three rates, i.e. 1 Mbps, 0.2 Mbps and 0.002 Mbps for UDP communication. Effect of changing the buffer size at the AP in terms of number of packets is also considered.

The rest of the paper is divided into three sections: Section 2 reviews the Key Hiding Scheme from the current perspective; Section 3 presents the considered network model including the simulation topology, parameters and performance metrics considered; Section 4 elaborates the results obtained with and without KHC; and Section 5 provides conclusion.

4.2 Review of Key Hiding Communication (KHC) Scheme

Key Hiding Communication (KHC) scheme [32] is a symmetric key-based secure communication method for WLANs where a new secret key is generated for encryption of each frame. The scheme proposes a key hiding concept for sharing and transferring the symmetric secret key and IV. Key hiding concept means that secret encryption key and IV are protected using counters and then mixed with each other such that key contents are not visible to the attacker and it becomes difficult for the attacker to get the key during communication. The mixed key and IV is termed as codeword (CD) and is added to the transmitted frame header. Upon receipt of frame at the receiver, key verification provides authentication per frame to the sender and also ensures possession of same key at the sender. The authentication involves lightweight operations like increment, XOR and modulus evaluations. MIC of only the authenticated frames is checked. The verified key is utilized to encrypt the data and evaluate MIC for the next frame. KHC hence follows the process of authenticating the sender and then utilizing the evolved key for providing security to next frame. Thus, KHC scheme provides per frame

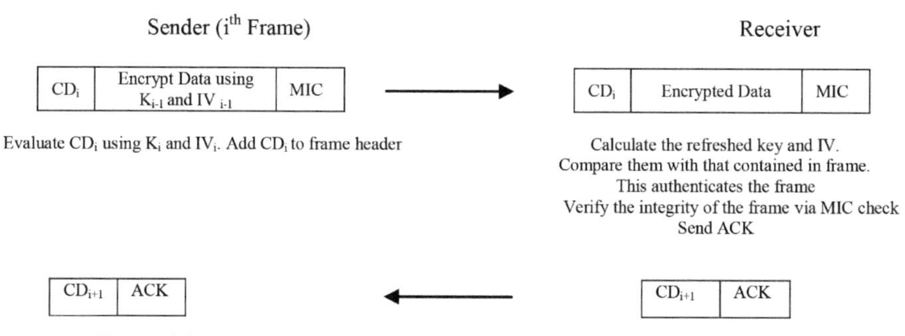

Figure 4.1 Review of KHC Scheme (Under Computational DoS Attack).

authentication, confidentiality and key freshness. The KHC communication pertaining to the current scope of work, i.e. under computational DoS attack is shown in Figure 4.1. The only time spent in the KHC scheme for per frame authentication is for key refreshing, i.e. hash evaluation. We assume that IV can be refreshed in parallel with key refreshing. Which means that after this time the AP is in position to verify the frame authenticity. As the time to verify the codeword is too less (only arithmetic operation like increment, XOR and MOD are used to verify codeword), the DoS attack on receiver conducted using frames with wrong codeword is not effective. Under attack, frames with correct codeword are easily identified and authenticated. On observing the loss of synchronization (when none of the received frame's codeword are correct), receiver requests the sender for resynchronization by sending acknowledgement (ACK) frame with header containing the current codeword and an encrypted value indicating index till where the KHC process was synchronized.

4.3 Network Model

Computational DoS attack on AP is studied using network simulator (NS2) [38–40]. Key Hiding Scheme is then tested for effectiveness under computational DoS attack on AP. Two processing delays of 0.02 ms and 0.2 ms at the AP are considered for cryptographic MIC evaluation and verification. The processing time of 0.2 ms is considered keeping in view of the results given in Ref. [30], where MIC calculation and verification takes time of the order of 0.193 ms on Acer Power Series PC (Processor: Core2Duo E6750@2.66 GHz ×2, RAM: 1GB, OS: Ubuntu 12.10). As the processing powers are increasing day by day, the processing time for MIC calculation and verification is also considered as 0.02 ms. Two different types of user traffic sources – File Transfer Protocol (FTP) and Constant Bit Rate (CBR) are considered. FTP uses Transmission Control Protocol (TCP) and tries to utilize the available bandwidth for communication via flow control. CBR uses User Datagram Protocol (UDP) and keeps on sending packets at the desired rate. Further, three CBR rates, i.e. 0.002 Mbps, 0.2 Mbps and 1 Mbps are considered for better understanding and evaluation purpose. For simulating a computation DoS attack three attacker nodes are considered. Each attacker node sends CBR traffic using UDP at 3 Mbps rate. This generates attack traffic at 9 Mbps rate which is sufficient in the considered WLAN scenario for the computational DoS attack at AP.

4.3.1 Simulation Topology

A WLAN scenario involving Wireless Station (STA) communicating via Access Point (AP) is considered as shown in Figure 4.2. AP is connected to the wired network having switch (S) and receiver nodes (R1 and R2). All the wired links are having 100 Mbps bandwidth. AP is having 11 Mbps bandwidth. Link AP-S is having 20 ms delay. Links S-R1 and S-R2 each have 2 ms delay. STA communicates with R1 while attacker nodes communicate with R2. Wireless station (STA) carry out two kinds of communication with receiver R1. First communication is done using File Transfer Protocol (FTP) over TCP and second communication is done using Constant Bit Rate (CBR) over UDP.

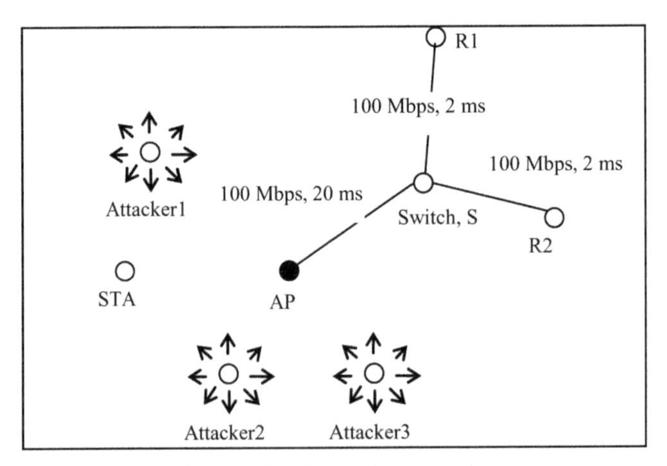

Figure 4.2 Simulation scenario.

For second communication, i.e. CBR over UDP, three sub cases with CBR rates as: 0.002 Mbps, 0.2 Mbps and 1 Mbps are considered. In all the cases, STA starts communication at 2.0 seconds while attacker nodes start communication at 10.0 seconds. Simulation duration for each run is 250 seconds.

4.3.2 Simulation Parameters

Table 4.1 lists the simulation parameters like number of nodes, their type, user application, protocols used, bandwidth details, link details, queuing strategy and simulation time. MAC layer parameters are also listed.

Table 4.1 Simulation parameters

S. No.	Simulation Parameter	Value(s)
1.	Number of wired nodes	3
2.	Number of wireless nodes	4 (1 Wireless STA while other 3 attackers)
3.	Number of attacker nodes	3 (attacker1, attacker2 and attacker3)
4.	User Applications	
	Constant Bit Rate (CBR)	0.002 Mbps, 0.2 Mbps and 1Mbps
	FTP	Available bandwidth
5.	Transport layer protocols	User STA: TCP and UDP
		Attacker: UDP
6.	AP Bandwidth	11 Mbps
7.	Wired link	Type, bandwidth, delay
	1. AP-S	Duplex, 100 Mbps, 20 ms
	2. S-STA2	Duplex, 100 Mbps, 2 ms
	3. S-STA3	Duplex, 100 Mbps, 2 ms
8.	Attacker node's attack rate	3 Mbps (using CBR over UDP)
9.	CBR packet size	1000 bytes
10.	Queuing strategy	Drop tail
11.	Simulation duration	250 s 2.0 s
	STA starts at:	10.0 s
	Attackers start at:	
12.	MAC layer parameters	
	Area	500 × 500 sq m
	SIFS	10 μs
	DIFS	50 μs
	Channel bit rate	11 Mbps
	Retry limit	7
	Slot time	20 μs
	Basic bit rate	2 Mbps
	MAC header	224 bits
	PHY header	192 bits

4.3.3 Performance Evaluation Metrics

For evaluating the KHC scheme under computational DoS, we consider two metrics defined as [37]:

1. Throughput: amount of data (in Bytes) received by destination per second.
2. End-to-End Delay: average amount of time taken in seconds by a TCP or UDP packet to travel successfully from source to destination.

4.4 Results and Discussion

The results are grouped in two parts: part I reports the results obtained when AP is under computational DoS attack without protection and part II reports the results obtained when AP is under computational DoS attack with KHC protection. For each part two cases are discussed. Case1, where processing delay per frame at the AP is considered as 0.02 ms and case2, where processing delay per frame at the AP is considered as 0.2 ms. For each of these, results for two user communication, i.e. FTP over TCP and CBR over UDP are shown. For second communication, i.e. CBR over UDP, three sub cases with CBR rates as: 0.002 Mbps, 0.2 Mbps and 1 Mbps are considered. In all the cases, wireless station's communication is affected only after the MAC buffer gets filled by the attack frames due to which the regular data frames are dropped. Results regarding throughput and delay for each of the cases are averaged over 30 simulation runs.

I. AP under computational DoS attack without Protection

CASE 1: Processing Time for MIC calculation and verification at AP = 0.02 ms

Communication using FTP over TCP:

Wireless station's transmission, i.e. FTP over TCP suffers under computational DoS attack conducted via wrong MIC verification. In this experiment, we consider time for MIC verification as 0.02 ms. The results are shown in Figure 4.3 (A and B), which clearly indicates that TCP throughput is almost reduced to negligible amount and the packets suffer large delay. The average throughput is 2 KB/s (2166.09 B/s) while average delay is 4.97 seconds.

Communication using CBR over UDP:

Constant bit rate (CBR) user traffic of the order of 0.002 Mbps, 0.2 Mbps and 1 Mbps are considered for the experiment. CBR uses UDP protocol for

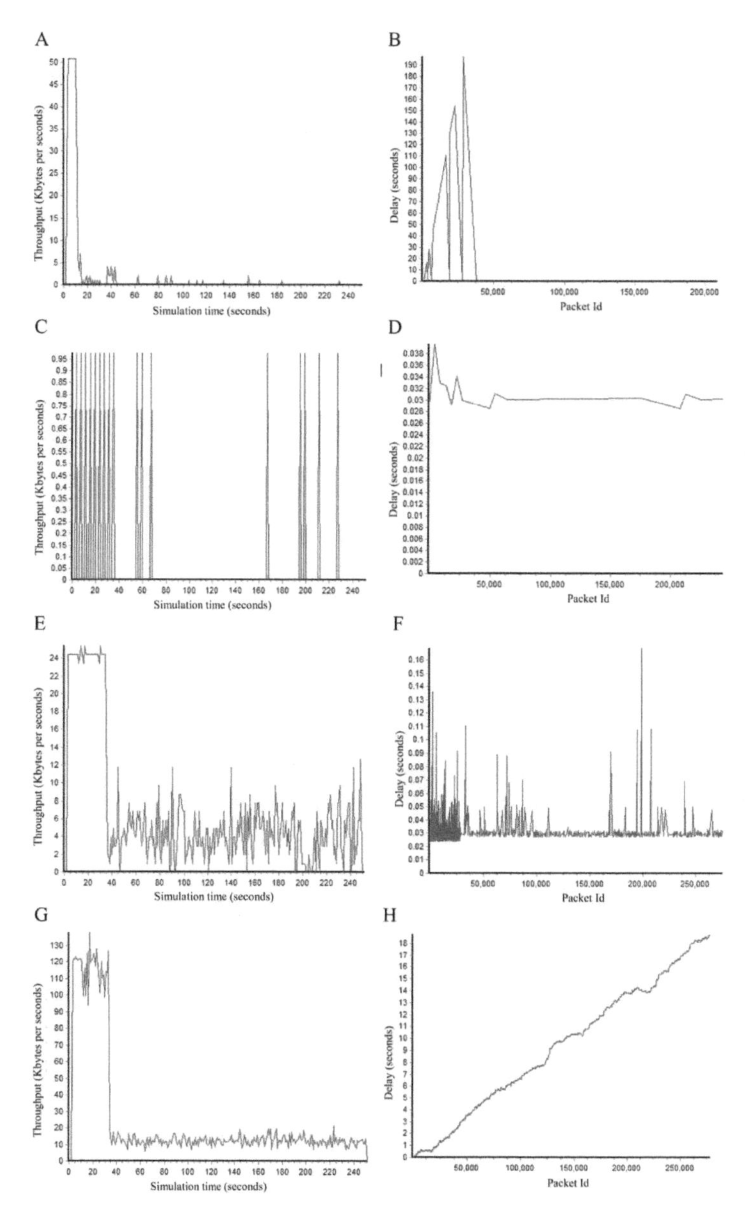

Figure 4.3 AP under 9 MB attack rate with processing time for MIC calculation and verification as 0.02 ms. (A) Throughput for FTP traffic over TCP (B) Delay for FTP over TCP (C) Throughput for 0.002 Mbps CBR traffic over UDP (D) Delay for 0.002 Mbps CBR traffic over UDP (E) Throughput for 0.2 Mbps CBR traffic over UDP (F) Delay for 0.2 Mbps CBR traffic over UDP (G) Throughput for 1 Mbps CBR traffic over UDP (H) Delay for 1 Mbps CBR traffic over UDP.

the data transfer. Several CBR rates are selected for analysis purpose. The results obtained are shown in Figure 4.3 (C–H). UDP traffic observes large packet drop in all the cases. At 0.002 Mbps CBR rate, a throughput of 76 B/s (75.56 B/s) while a delay of 0.0305 seconds is observed. At 0.2 Mbps CBR rate, a throughput of 7 KB/s (7016.19 B/s) while a delay of 0.0305 seconds is observed. At 1 Mbps CBR rate a throughput of 25 KB/s (26052.42 B/s) while a delay of 4.672 seconds is observed. It is hence clear from the results that the UDP-based user traffic is affected under the computational DoS attack.

CASE 2: Processing Time for MIC calculation and verification at AP = 0.2 ms

Communication using FTP over TCP:

Wireless station's transmission, i.e. FTP over TCP suffers under computational DoS attack conducted via wrong MIC verification. We consider time for MIC verification as 0.02 ms. Results are shown in Figure 4.4 (A and B) which clearly indicate that TCP throughput is almost reduced to negligible amount and the packets suffer large delay. The average throughput is 2 KB/s (2129.27 B/s) while average delay is 6.57 seconds.

Communication using CBR over UDP:

Three Constant Bit Rate (CBR) user traffic of 0.002 Mbps, 0.2 Mbps and 1 Mbps rates are considered for the experiment. The results obtained are shown in Figure 4.4 (C–H). UDP traffic observes large packet drop in all the cases. At 0.002 Mbps CBR rate, a throughput of 33.3 B/s (33.3B/s) while a delay of 0.293 seconds is observed. At 0.2 Mbps CBR rate, a throughput of 1 KB/s (1048.78 B/s) while a delay of 42.31 seconds is observed. At 1 Mbps CBR rate a throughput of 4 KB/s (4321.14 B/s) while a delay of 103.189 seconds is observed. It is hence clear from the results that UDP-based user traffic is badly affected under the computational DoS attacks involving processing at AP of 0.2 ms.

In all these simulations we consider the buffer size at the AP as 10,000 packets. If we reduce the buffer size to 100 and consider the TCP communication case then the average throughput was somewhat improved. For example, in 0.02 ms processing case throughput is around 36 KB/s (36923.33 B/s) with delay of 0.387 seconds. This improvement is due to the fact that because of small size buffer the attack packets get dropped and processing is not involved for these packets which give more chances to the packets of wireless

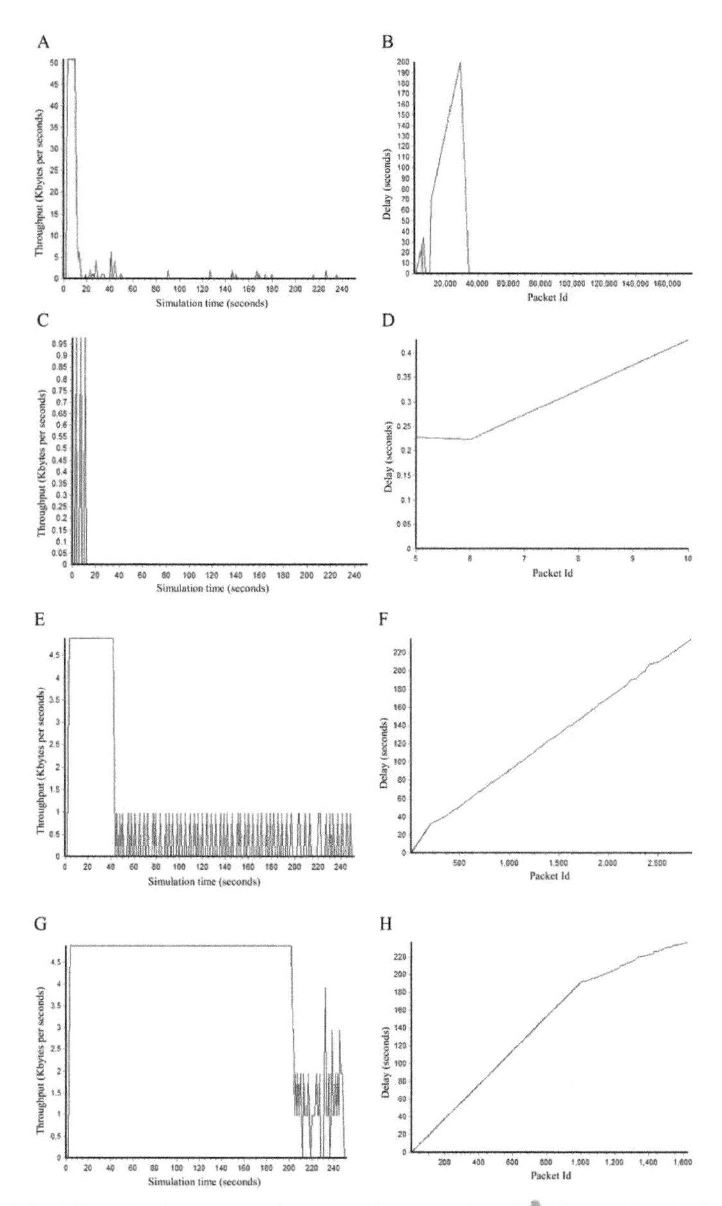

Figure 4.4 AP under 9 MB attack rate with processing time for MIC calculation and verification as 0.2 ms. (A) Throughput for FTP traffic over TCP (B) Delay for FTP traffic over TCP (C) Throughput for 0.002 Mbps CBR traffic over UDP (D) Delay for 0.002 Mbps CBR traffic over UDP (E) Throughput for 0.2 Mbps CBR traffic over UDP (F) Delay for 0.2 Mbps CBR traffic over UDP (G) Throughput for 1 Mbps CBR traffic over UDP (H) Delay for 1 Mbps CBR traffic over UDP.

station. If we consider the UDP communication case with reduced buffer size then the average throughput gets low. For example, in 0.02 ms processing case with 0.002 Mbps data rate, throughput is around 23 B/s (23.47 B/s) while delay is around 0.0316 seconds. This deterioration is due to the fact that because of small size buffer the packets of wireless station also get dropped.

II. AP protected by KHC scheme from computational DoS attack

CASE 1: Processing Time for MIC calculation and verification at AP = 0.02 ms

Communication using FTP over TCP.

User communication using FTP over TCP transmission is considered with KHC protection at AP under computational DoS attacks. In this experiment the MIC computation and verification delay is considered as 0.02 ms. Results are shown in Figure 4.5 (A and B). The average throughput is 82 KB/s (84432.29 B/s) which is much greater than the throughput of TCP communication without KHC, i.e. 2 KB/s. The average delay is 0.0585 seconds which is less as compared to the delay when AP is not under protection. i.e. 4.971 seconds.

CBR over UDP communication.

User communication using CBR over UDP transmission is considered with KHC protection at AP under computational DoS attacks. User traffic of the order of 0.002 Mbps, 0.2 Mbps and 1 Mbps are considered in this experiment. The results are shown in Figure 4.5 (C–H). At 0.002 Mbps CBR rate, a throughput of 253 B/s (253.01 B/s) while a delay of 0.0323 seconds is observed. At 0.2 Mbps CBR rate, a throughput of 24 KB/s (24907.63 B/s) while a delay of 0.0318 seconds is observed. At 1 Mbps CBR rate, a throughput of 112 KB/s (115132.53 B/s) while a delay of 9.086 seconds is observed. The results obtained at 0.002 Mbps and 0.2 Mbps are indicative that almost all the STA traffic is successfully transferred while maintaining appropriate delay. The results obtained at 1 Mbps are indicative of the MAC layer contention between the STA frames and attack frames. Due to this contention, per frame delay increases as most of the time the STA's frames are waiting in the interface queue for the free channel. The high throughput indicates effectiveness of the KHC protection at the AP under computational DoS attacks involving processing of 0.02 ms.

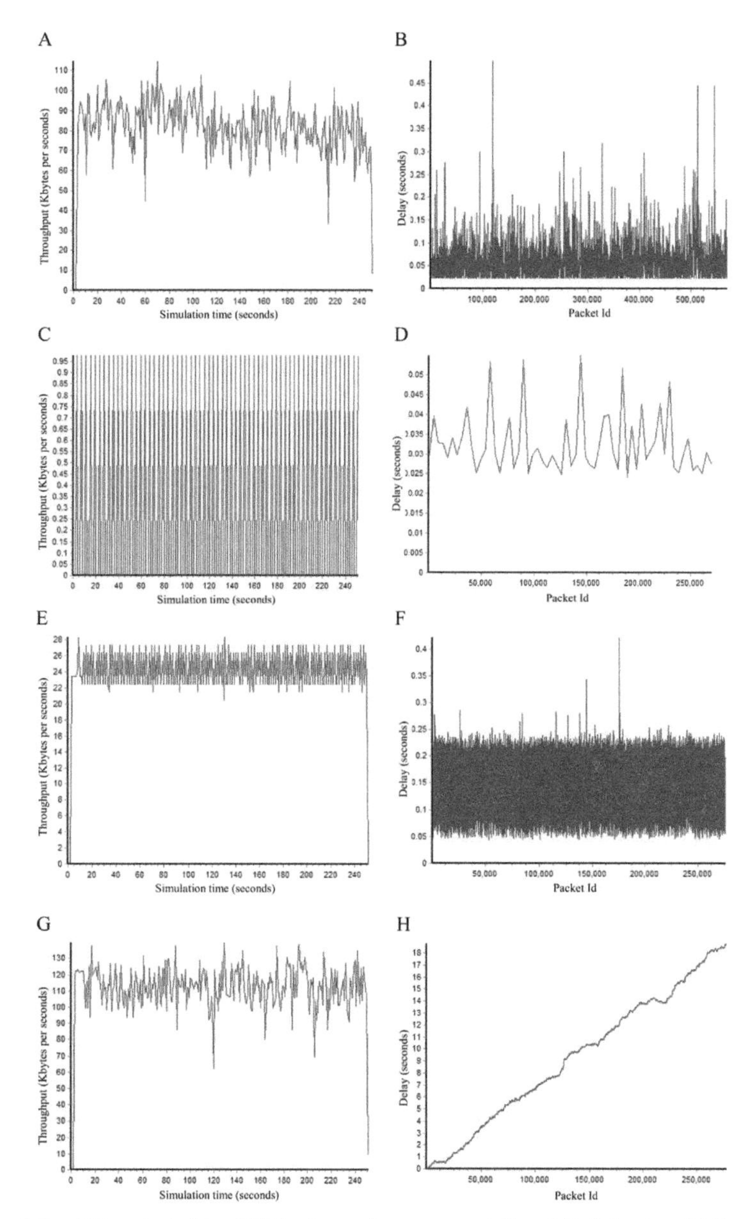

Figure 4.5 AP protected by KHC scheme from computational DoS having 9 Mbps attack rate with processing time for MIC calculation and verification as 0.02 ms. (A) Throughput for FTP traffic over TCP (B) Delay for FTP over TCP (C) Throughput for 0.002 Mbps CBR traffic over UDP (D) Delay for 0.002 Mbps CBR traffic over UDP (E) Throughput for 0.2 Mbps CBR traffic over UDP (F) Delay for 0.2 Mbps CBR traffic over UDP (G) Throughput for 1 Mbps CBR traffic over UDP (H) Delay for 1 Mbps CBR traffic over UDP.

CASE 2: Processing Time for MIC calculation and verification at AP = 0.2 ms

Communication using FTP over TCP:

In this experiment, STA communication using FTP over TCP transmission is considered with KHC protection at AP under computational DoS attacks. The MIC computation and verification delay is considered as 0.02 ms. Results are shown in Figure 4.6 (A and B). The average throughput is 75 KB/s (76609.32 B/s) which is much greater than the throughput of TCP communication without KHC, i.e. 2 KB/s. The average delay is 0.167 seconds which is less as compared to the delay when AP is not under protection, i.e. 6.57 seconds. The average throughput is less and delay here is more as compared with the previous case, i.e. FTP over TCP communication with 0.002 ms processing time. Due to the increased delay less number of packets are sent as compared with the FTP over TCP communication with 0.002 ms processing time.

CBR over UDP communication:

STA communication using CBR over UDP transmission is considered with KHC protection at AP under computational DoS attacks. User traffic of the order of 0.002 Mbps, 0.2 Mbps and 1 Mbps are considered in this experiment. The results are shown in Figure 4.6 (C–H). At 0.002 Mbps CBR rate, a throughput of 253 B/s (253.06 B/s) while a delay of 0.14 seconds is observed. At 0.2 Mbps CBR rate, a throughput of 24 KB/s (24987.90 B/s) while a delay of 0.144 seconds is observed. At 1 Mbps CBR rate, a throughput of 112 KB/s (115100.40 B/s) while a delay of 9.196 seconds is observed. The results obtained at 0.002 Mbps and 0.2 Mbps are indicative that almost all the STA traffic is successfully transferred while maintaining appropriate delay. The results obtained at 1 Mbps are indicative of the MAC layer contention between the STA frames and attack frames. Due to this contention, per frame delay increases as most of the time the STA's frames are waiting in the interface queue for the free channel. The high throughput indicates the effectiveness of the KHC protection at the AP. Thus, through simulation experiments effectiveness of the KHC under computational DoS attack is established.

For KHC protected cases, the frame buffer size at AP is also varied as 50, 100, 200, 500, 1000 and 10,000 frames but this change shows no effect on the scheme as the average buffer length observed during the simulation is around 2.7 packets.

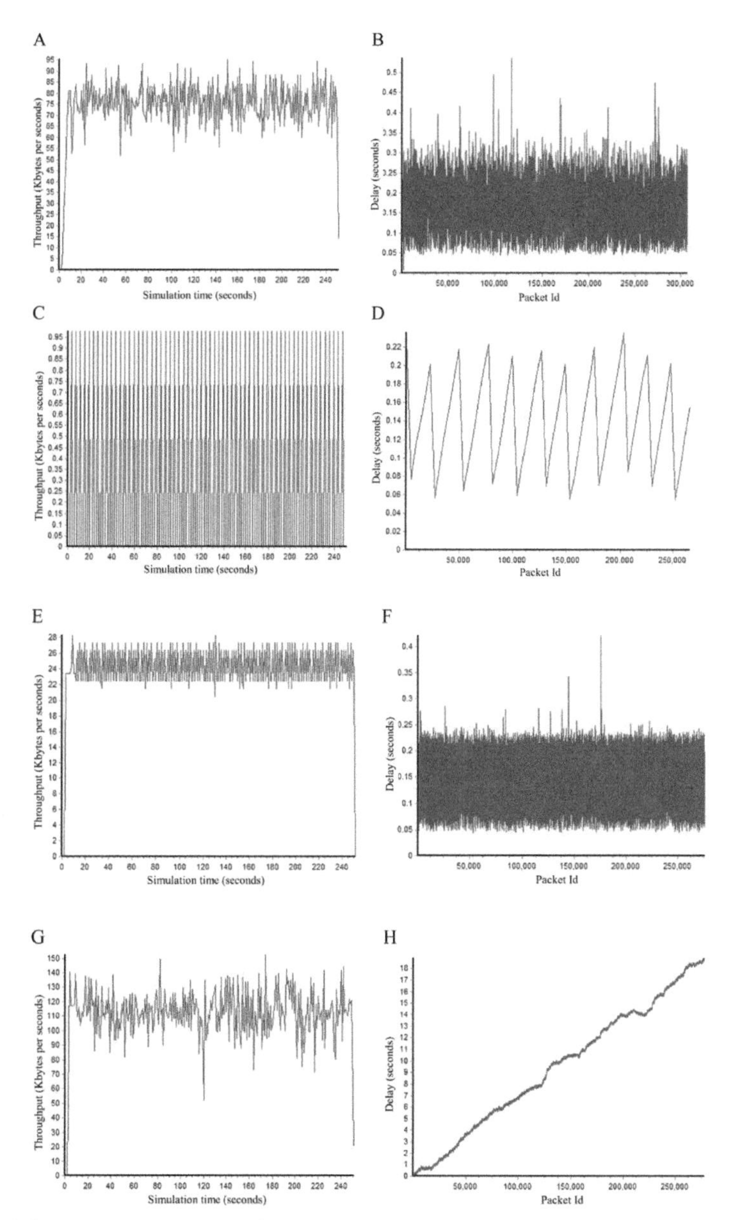

Figure 4.6 AP protected by KHC scheme from computational DoS having 9 Mbps attack rate with processing time for MIC calculation and verification as 0.2 ms. (A) Throughput for FTP traffic over TCP (B) Delay for FTP over TCP (C) Throughput for 0.002 Mbps CBR traffic over UDP (D) Delay for 0.002 Mbps CBR traffic over UDP (E) Throughput for 0.2 Mbps CBR traffic over UDP (F) Delay for 0.2 Mbps CBR traffic over UDP (G) Throughput for 1 Mbps CBR traffic over UDP (H) Delay for 1 Mbps CBR traffic over UDP.

4.5 Conclusion

Most types of DoS attacks in WLANs are performed using either control or management frames. We show that even data frames can be used for this purpose. Main contribution of the paper is establishment of the processing based model at AP in NS2. NS2 model currently does not involve processing at a wireless node during simulation. This restricts the use of NS2 for cryptographic-based protocol validation. In the current NS2 model, the processing delay was added for each packet that AP receives. This processing delay method developed can even be used for analysing the cryptographic protocol's behaviour at AP and also on the entire network. We selected two processing times, i.e. 0.02 ms and 0.2 ms for MIC calculation and verification at the AP. Processing time of 0.2 ms at AP for MIC verification is selected keeping the current CPU processing powers while processing time of 0.02 ms at AP for MIC calculation and verification is selected keeping in view the new upcoming fast systems.

Using the processing delay model, computational flooding DoS attack on AP is realized. DoS attack effect increases with increase in the computational time. We find major reason that attributed to success of the computational DoS attack as consumption and wastage of AP resources where the attack packets not only occupy the AP's buffer space but also utilize the processing power of AP. The results show that AP suffers under computational DoS attack when no protection is provided to it. For the entire duration of flooding DoS attack, regular STA communication is affected severely.

For showing effectiveness of the KHC scheme an access point is considered under computational DoS attacks. The AP protected using KHC scheme performs efficiently under the computational DoS attack. This clearly indicates the effectiveness of the KHC scheme under computational DoS attack. We analyse the behaviour of KHC under DoS attack not only for the current systems but also for the new and improved systems having higher processing power.

Two of the major findings of the paper are: first, establishment of the fact that processing at AP may lead to DoS attack and second, validation of KHC scheme against computational DoS attacks. For reducing the processing overheads at AP, authentication and integrity measures involved per frame should be made lightweight in nature. Several lightweight authentication solutions are proposed by researchers [41–48] that not only use less number of bits for authentication but also involve less processing time. These are less secure. Hence, efforts should be done to enhance security procedures in these lightweight authentication mechanisms so that they provide per frame authentication and integrity equivalent to message integrity code. Thus, in future for protecting against flooding DoS attack, the upcoming solutions should not only try to propose for utilizing the MAC buffer appropriately but should also work to reduce processing at the AP.

References

[1] Bellardo, J., Savage, S. (2003). "802.11 denial-of-service attacks: real vulnerabilities and practical solutions," in *Proceedings of the 12th Conference on USENIX Security Symposium*, Washington, DC, 15–28.

[2] Bernaschi, M., Ferreri, F., Valcamonici, L. (2008). Access points vulnerabilities to DoS attacks in 802.11 networks. *Wireless Netw.* 14, 159–169.

[3] Bicakci, K., Tavli, B. (2009). Denial-of-service attacks and countermeasures in IEEE 802.11 wireless networks. *Elsevier Comput. Stds. Inter.* 31, 931–941.

[4] Pelechrinis, K., Iliofotou, M., Krishnamurthy, S. V. (2011). Denial of service attacks in wireless networks: the case of jammers. *IEEE Commun. Surv. Tutor.* 13, 245–257.

[5] Dsniff (2016). *Collection of Tools for Network Penetration "Dsniff"*. Available at: http://packages.debian.org/stable/net/dsniff/

[6] GNU MAC (2017). *Changer 16.0*. Available at: http://www.alobbs.com

[7] SpoofMAC (2000). MAC Spoofing "SpoofMAC". Available at: http://www.klcconsulting.net/smac/

[8] Joshua, W. (2015). File2air written by Joshua Wright. Available at: http://www.willhackforsushi.com/File2air.html

[9] Aireplay (2017). *Frame Injection Program "Aireplay-ng"*. Available at: http://www.aircrack-ng.org/doku.php

[10] Airjack (2016). *Rogue DoS Attack Tools by Forging Management Frames*. "Airjack". Available at: http://sourceforge.net/projects/airjack/

[11] Void11 (2016). *WLAN DoS Attack Tool* "Void11". Available at: http://www.wlsec.net/void11/

[12] Aircrack-ng (2017). *WPA-PSK Keys Cracking Program "Aircrack-ng"*. Available at: http://www.aircrack-ng.org/

[13] Packet forge-ng (2006). *Encrypted Packets Injection Program "Packet forge-ng"*. Available at: http://www.aircrack-ng.org/doku.php

[14] IEEE 802.11i (2004). *Wireless LAN Medium Access Control (MAC) and Physical Layer (PHY) Specifications: Medium Access Control (MAC) Security Enhancements.* Rome: IEEE Standard.

[15] He, C., and Mitchell, J. C. (2005). "Security analysis and improvements for IEEE 802.11i," in *Proceedings of the Annual Network and District System Section Symposium (NDSS'05)*, Porto Alegrem, 90–110.

[16] Al Naamany, A. M., Shidhani, A. A., Bourdoucen, H. (2006). IEEE Wireless LAN Security. *Int. J. Comput. Sci. Netw. Sec.* 6, 5B.

[17] NIST Special Publication 800-97. (2007). *Establishing Wireless Robust Security Networks: A Guide to IEEE 802.11i.* Gaithersburg, MD: NIST Special Publication.

[18] Xing, X., Shakshuki, E., Benoit, D., Sheltami, T. (2008). "Security analysis and authentication improvement for IEEE 802," in *Proceedings of the 11i Specification. in IEEE GLOBECOM 2008 Global Telecommunications Conference*, Miami, FL, 1–5.

[19] Prodanovic, R., and Simic, D. (2007). A survey of wireless security. *J. Comput. Inf. Technol.* 15, 237–255.

[20] Turab, N., Moldoveanu, F. (2009). A comparison between Wireless LAN security protocols. U.P.B. *Sci. Bull. Series C* 71, 1.

[21] Denial of Service Attacks and Mitigation (2016). Available at: http://www.sans.org/reading_room/whitepapers/wireless/80211-denial-service-attacks-mitigation_2108 802.11

[22] Malekzadeh, A., Ghani, A. A., and Subramaniam, S. (2011). Protected control packets to prevent denial of services attacks in IEEE 802.11 wireless networks. *EURASIP J. Inf. Secur.* 4, 1–20.

[23] Malekzadeh, M., Ghani, A. A., and Subramaniam, S. (2011). Design and implementation of a lightweight security model to prevent IEEE 802.11Wireless DoS Attacks. *EURASIP J. Wire. Commun. Netw.*, 1–16.

[24] Malekzadeh, M., Ghani, A. A., Zulkarnain, Z. A., and Muda, Z. (2007). Security improvement for management frames in IEEE 802.11 wireless networks. *Int. J. Comput. Sci. Netw. Secur.* 7, 276–284.

[25] Liu, C., and Yu, J. (2007). A solution to WLAN authentication and association DoS attacks. *Int. J. Comput. Sci.* 34, 31–36.

[26] Ding, P., Holliday, J., Celik, A. (2004). Improving the security Wireless LANs by managing 802.1X disassociation. *Consum. Commun. Netw. Conf.*, 53–58.

[27] Aslam, B., Islam, M. H., Khan, S. A. (2006). Pseudo randomized sequence number based solution to 802.11 disassociation denial of service attack, *Intl. Conf. Mobile Comput. Wirel. Commn.*, 215–220.

[28] Aslam, B., Islam, M. H., Khan, SA. (2006). 802.11 disassociation DoS attack and its solutions: a survey. *Intl. Conf. Mobile Comput. Wireless Commun.*, 221–226.

[29] Lockhart, A. (2005). *Deauthentication Frame DoS, 2005*. Available at: http://www.wirelessve.org/entries/show/WVE-2005-0045

[30] Singh, R., and Sharma, T. P. (2013). A secure WLAN authentication scheme, IEEK. *Trans. Smart Process. Comput.* 2, 176–187.

[31] Singh, R., and Sharma, T. P. (2011). "Detecting and reducing denial of service attacks in WLANs," in *Proceedings of the World Congress On Information and Communication Technologies*, 968–973.

[32] Singh, R., and Sharma, T. P. (2013). A key hiding communication scheme for enhancing the Wireless LAN Security. *Wireless Pers. Commun.* 77, 1145–1165. doi: 10.1007/s11277-013-1559-0.

[33] Holt, A., and Huang, C. Y. (2010). *802.11 Wireless Networks: Security and Analysis*. Berlin: Springer-Verlag, 2010.

[34] Martinovic, I., Zdarsky, F. A., Bachorek, A., Schmitt, J. B. (2007). "Measurement and Analysis of Handover Latencies in IEEE 802.11i Secured Networks," in *Proceedings of the European Wirel. Conference (EW2007)*, Paris, 1–7.

[35] Martinovic, I., Zdarsky, F. A., Bachorek, A., Schmitt, J. B. (2006). *Introtuction of IEEE 802.11i and Measuring its Section vs Performance Tradeoff.* Technical Report 351/06. Kaiserslautern: University of Kaiserslautern.

[36] Suleiman, K. H., Javidi, T., Liu, M., and Kittipiyakul, S. (2008). *The Impact of MAC Buffer Size on the Throughput Performance of IEEE 802.11*. Technical Report, 2008. Kaiserslautern: University of Kaiserslautern.

[37] Malekzadeh, M., Azim, A., Ghani, A., Subramaniam, S., and Desa, J. M. (2011). Validating Reliability of OMNeT++ in Wireless Networks DoS Attacks: Simulation vs. Testbed. *Int. J. Netw. Secur.* 13, 13–21.

[38] Building Ns (2005). Available at: http://www.isi.edu/nsnam/ns/ns-build.html

[39] The VINT Project (2011). Available at: http://www.isi.edu/nsnam/ns/ns_doc.pdf

[40] Issariyakul, T., Hossain, E. (2009). *Introduction to Network Simulator NS2*. Berlin: Springer.

[41] Johnson, H., Nilsson, A., Fu, J., Wu, S. F., Chen, A., Huang, H. (2002). "SOLA: a one bit identity authentication protocol for access

control in IEEE 802.11," in *Proceedings of the IEEE Global Telecommns Conference, GLOBECOM'02*, San Francisco, CA, 768–772.

[42] Wu, F., Jonson, H., Nilson, A. (2004). SOLA: Lightweight Security for Access Control in IEEE 802.11. *Wireless Sec.* 2004, 10–16.

[43] Wang, H., Velayutham, A., Guan, Y. (2003). "A Lightweight Authentication Protocol for Access Control in IEEE 802.11," in *Proceedings of the IEEE Global Telecommunications Conference, GLOBECOM'03*, San Francisco, CA, 1384–1388.

[44] Wang, H., Cardo, J., Guan, Y. (2005). Shepherd: a lightweight statistical authentication protocol for access control in wireless LANs. *Comput. Commun.* 28, 1618–1630.

[45] Ren, K., Lee, H., Park, J., Kim, K. (2004). "An enhanced lightweight authentication protocol for access control in Wireless LANs," in *Proceedings of the 4th International Conference on Networks, ICON'04*, Daejeon, 444–450.

[46] Lee, Y.-S., Chien, H.-T., Tsai, W.-N. (2009). Using Random Bit Authentication to defend IEEE 802.11 DoS attacks. *J. Inf. Sci. Eng.* 25, 1485–1500.

[47] Pepyne, D. L., Ho, Y.-C., Zheng, Q. (2003). SPRiNG: Synchronized random numbers for wireless security. in *Proceedings of the IEEE Wireless Communications and Networking, WCNC'03*, Las Vegas, NV, 2027–2032.

[48] Lee, I. (2010). A *Novel Design and Implementation of DoS-Resistant Authentication and Seamless Handoff Scheme for Enterprise WLANs.* M.Tech. thesis, Department of Computer Science and Software Engineering, University of Canterbury, Christchurch.

5

Development of Computation Algorithm and Ranking Methods for Decision-Making under Uncertainty

Alexander V. Bochkov and Nikolay N. Zhigirev

Risk Analysis Center, NIIgazeconomika, LLC, Moscow, Russia

This chapter presents the two methods of ranking objects used to make a decision if the situation is unclear. The first part of this chapter examines the method of putting the objects arranged within homogeneous groups in accordance with a specified criterion into a general ranked list. This task is often performed in the process of decision-making, and it includes ranking diverse objects of complex technical systems, comparing the ratings of countries and companies made by different rating agencies, etc. The proposed algorithm completes the task with the help of partial expert comparisons between the pairs of objects that belong to different ranked lists. A theoretical justification and a practical example of calculations are given in the first part. The second part observes the method that helps reconstruct the preferences experts had as far as the method of analysing hierarchies is concerned if for some reason experts could not compare the pairs of objects. We examined the most common decision-making process because of its efficiency, flexibility and simplicity – the method of pair congruences. The main disadvantage of this method when examining a large number of alternatives or in a rather wide field of knowledge is the impossibility of comparing each element with each, both because of the large number of such comparisons, random passes, and because of the expert's difficulties when comparing some alternatives. In the estimates there are missing data, making it difficult to make a decision, because most statistical methods are not applicable to an incomplete set of data. We cannot work with a matrix that contains mostly zero elements and a rather popular algorithm for processing matrices of pairwise comparisons (the Saaty algorithm). The purpose of the article is to develop a method for processing incomplete comparison matrices in order to obtain weights (coefficients) of the considered alternatives, which allow us to compare them quantitatively.

Methods. In practice, there are several approaches to work with data arrays that contain missing values. The first approach, the easiest to implement, is to delete instances that contain missing values from the array and work only with full data. The use of this approach is advisable if the data gaps are single. But even in this case there is a serious danger in the removal of data to "lose" important regularities. The second approach is the use of special modifications of data processing methods that allow the presence of omissions in the array. Finally, various methods are used to estimate the values of missing elements. These methods help fill-in gaps in arrays, based on some assumptions about the significance of missing data. The principle applicability and effectiveness of an approach depend on the number of omissions in the data and the reasons for which they were formed. In the article, the matrix of pairwise comparisons is considered in the format of a loaded graph, and alternatives are vertices, and comparisons between them are edges of the graph. Accordingly, if a pair of alternatives arises, for which the expert could not specify a preference, then the corresponding edge is absent. The method of removing edges corresponding to the most contradictory estimates is considered. Algorithm for breaking the cycles, leading to the transformation of the original graph to the spanning tree, makes it possible to unequivocally compare any two alternatives. The algorithm of joint agreement between the upper and lower bounds of expert assessments is not considered in this article.

Results. The article gives an example of the practical application of the developed algorithm for processing an incomplete matrix of pairwise comparisons of ten objects obtained during a certain examination. The working capacity of the proposed approach to the tasks of restoring the priorities of the compared alternatives is shown, the ways of calculating automation and the direction of further research are outlined.

Conclusions. The proposed method can be applied to a wide range of tasks of analysis and quantification of risks, management of the security of complex systems and facilities, as well as tasks related to monitoring compliance with requirements for such highly reliable elements as elements of nuclear reactors, aviation and rocket and space equipment, gas equipment and the like, i.e. where it is required to estimate the small (less than 0.01) probability of failure for a given operating time, and the failure statistics of such elements in operation are practically absent. The proposed algorithm can be used for expert evaluation to establish the type and parameters of the distribution of the operating time to failure of such highly reliable elements, which in turn will allow us to measure reliability with acceptable accuracy.

5.1 Trough-Ranking Method for a Regulate Lists Objects of Different Types by Partial Expert Comparisons

5.1.1 Literature Review

One-type object ranking with respect to the given criterion is not difficult if there is background information about resource criteria and basic ones. One-type objects carry out similar tasks. Qualitative differences, if existent, characterize the way they function, which is taken into account when it comes to individual techniques for measuring their significance. Moreover, similar parameters are used to describe one-type objects, and this fact guarantees higher level of authenticity of the received comparative estimates.

The task of ranking occurs quite often. The objects ranking methods are based on mathematical simulation, expert reports, decision-making theory, and interval estimation [1–3]. At the same time currently available ranking methods (for instance, ranking of the facilities by the extent of their protection in case of emergencies on railroad transport [4], ranking of the hazardous production facilities of the gas distribution systems [5], etc.) fail to take into account the properties of the structural connectivity of the ranking objects and importance of the specific facility operation for interfacing systems and subsystems.

The task of objects ranking is a standard issue for the theory of measuring some complex synthetic features of the facilities [6]. A formal result is obtained by way of plotting some validity or usefulness function that links the measured feature with simpler resource indicators (factors) measured in actual values [7]. The value function is used to settle the issues of selecting the best variant from the set of alternatives [8], and also composition issues, e.g., the issue of forming the portfolio of orders for works, provided the resources are limited (volume of financing the creation or modification of the facilities [9]. The factors used to obtain the ranks are often measured in qualitative scales rather than in quantitative ones, which results in use of expert evaluation methods and expert system technologies to construe relationships between utility and primary resource factors [10, 11]. Development of computer engineering makes it possible to assess the facilities with description factors set with an error. This necessitates development of a specific apparatus for statistical processing of the primary data [12] and use of the fuzzy logic tools [9, 13–15].

Solution of the ranking issues is characterized by an adaptive nature of the decision-making procedure to be followed to select optimal variants [14, 16], under which several experimental data and expert preferences correlation cycles for development of the final formula [17] must be performed.

In order to solve a mathematically nontrivial problem of putting the given lists into a general list that ranks every object by its significance regardless of its type, a number of original innovative approaches are to be taken. They ensure the correct comparison of the objects of various types on the basis of partial expert conclusions which reveal that the distinguished representative pairs of the objects of various types are equivalent.

5.1.2 Algorithm Description

We shall pay more attention to the description of the relevant algorithm. It is worth noting that estimating objects' state and their significance is a nontrivial task. The first thing one has to do is to complete it to measure both qualitative and quantitative characteristics that are sufficient for any further estimates. Since the measurements can be made with a certain precision, the list of indicators that suggest the state of objects is always conditionally sufficient because it cannot fully embrace a wide range of internal and external circumstances which determine the function of the objects. That said, one can only speak about the measurements and estimates with certain errors.

In fact, any estimate is to be treated as an estimate of the position of a dynamic object indicator value close to its equilibrium (Figure 5.1).

The equilibrium value practically never equals to the estimate value but the estimate is always in the domain of attraction (DA) of the equilibrium point, with the dimension of $\Delta x_{abs}\,(\mathrm{O})$ (in case of linear models of objects O), proportional to the measured objects DA $x\,(\mathrm{O})$.

$$\Delta x_{abs}\,(\mathrm{O}) = \delta_{abs} \cdot x\,(\mathrm{O})\,. \tag{5.1}$$

When it comes to the DA size, classical measurement theories state that it is regarded as absolute measurement error $\Delta x_{abs}\,(\mathrm{O})$. In order to make the

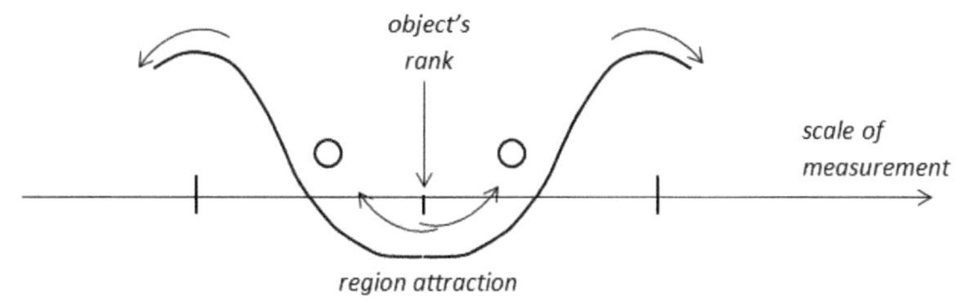

Figure 5.1 The position of a dynamic object indicator value close to its equilibrium.

characteristics of measurement errors independent on the size of an object, the logarithms are quite often taken of the measurements, thus estimating the domain of attraction by a relative error on a logarithmic scale.

$$\delta_{rel} = \log\left(1 + \delta_{abs}\right). \tag{5.2}$$

In the formulae (5.1) and (5.2) the value δ_{abs} is often measured as a percentage of the estimate value $x\left(\mathrm{O}\right)$. The object O is believed to change its state or one deals with the estimates of another object O′ if $x\left(\mathrm{O}'\right) - x\left(\mathrm{O}\right) > \delta_{abs} \cdot x\left(\mathrm{O}\right)$ or

$$\left|\log\left(x\left(\mathrm{O}'\right)\right) - \log\left(x\left(\mathrm{O}\right)\right)\right| > \delta_{rel}. \tag{5.3}$$

Additionally, δ_{rel} (in linear object models) is already independent on the size which facilitates any further constructions.

Proceeding from Equation (5.3) we shall consider the object O_1 to be more valuable than the object O_2 if

$$\log\left(x\left(\mathrm{O}_1\right)\right) > \log\left(x\left(\mathrm{O}_2\right)\right) + \delta_{rel}. \tag{5.4}$$

The case when the condition (5.3) is not met, requires further consideration. As it was mentioned above, as a rule, the self-assessment of the object O is not equal to the estimate made with relation to the same object O in different time.

Nevertheless, we regard them as the estimates of one and the same object O since the position of the DA boundaries is stable. Consequently, we can record the notion of «approximate equality» and formally write it down in the following way:

$$\mathrm{O} \approx \mathrm{O}. \tag{5.5}$$

Apparently, if one takes two arbitrary points of measurement time, the inequation $\left|\log x\left(\mathrm{O}\left(t_1\right)\right) - \log x\left(\mathrm{O}\left(t_2\right)\right)\right| < 2 \cdot \delta_{rel}$ will be always satisfied. The inequations

$$\log x\left(\mathrm{O}\left(t_2\right)\right) > \log x\left(\mathrm{O}\left(t_1\right)\right) \tag{5.6}$$

$$\log x\left(\mathrm{O}\left(t_2\right)\right) < \log x\left(\mathrm{O}\left(t_1\right)\right) \tag{5.7}$$

will be equiprobably satisfied (50%).

Figure 5.2 illustrates the comparison of the objects O_1 and O_2 of the same type.

Let us review in detail the cases of comparison shown in Figure 5.2.

Case 1. $\log x\left(\mathrm{O}_1\right) = \log x\left(\mathrm{O}_2\right)$ – the geometrical arrangement showcases the fifty-fifty situation.

Case 2. $\log x\left(\mathrm{O'}_1\right) > 2 \cdot \delta_{rel} + \log x\left(\mathrm{O}_2\right)$ – the first object is estimated higher than the other. It undoubtedly belongs to another domain of attraction, and it is undoubtedly more significant. In this case we shall design it as $\mathrm{O}_1 > \mathrm{O}_2$. Figure 5.3 demonstrates a special case of comparison.

Case 3. $\log x\left(\mathrm{O''}_1\right) = \log x\left(\mathrm{O}_2\right) + \xi,\ \xi < 2\delta$. The probability p_{12} of significance $\mathrm{O''}_1$, which is higher than significance O_2 is shifted from 0.5 to 1 under the square law

$$p_{12} = 1 - \left(\frac{2\delta - \xi}{2\delta}\right)^2, \tag{5.8}$$

but it is still not equal to one.

In this case we shall continue to believe that $\mathrm{O''}_1 \sim \mathrm{O}_2$. In this case we shall continue to believe that .... since there are no «sound data» which can

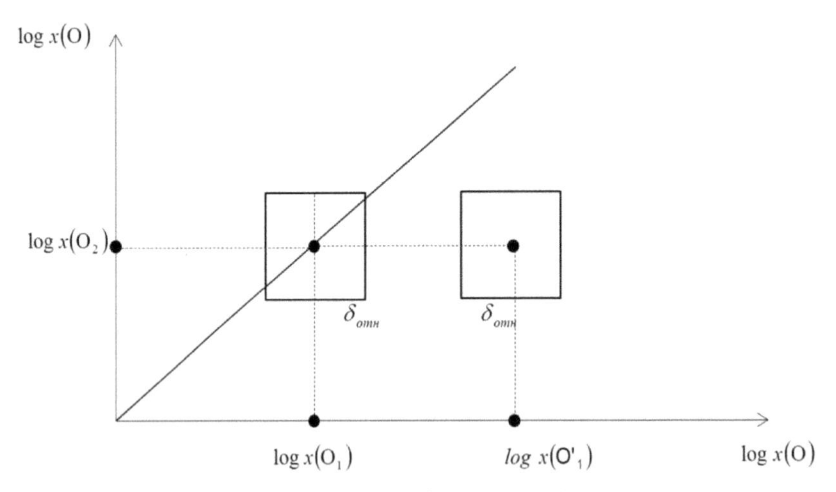

Figure 5.2 The comparison of the objects of the same type.

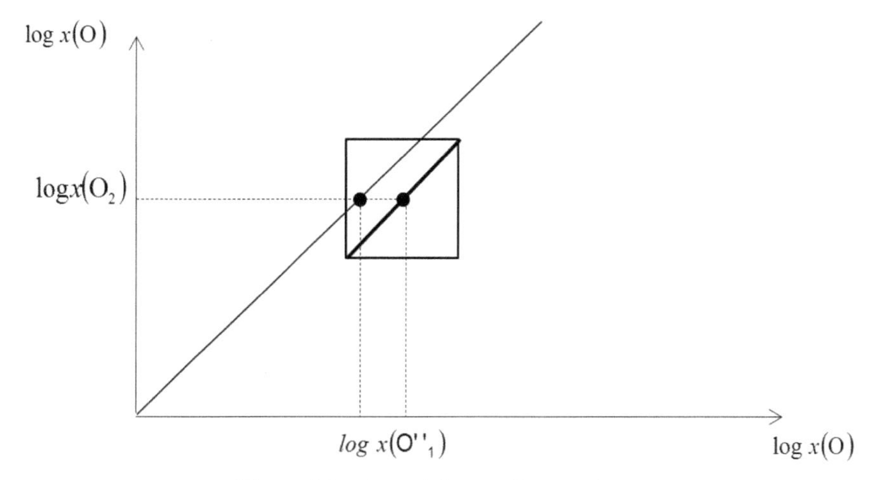

Figure 5.3 A special case of comparison.

prove if O''_1 is more significant than O_2 since there is only one measurement for each one, and $\log x\,(O''_1)$, and $\log x\,(O_2)$ are calculated with the help of the same calculation scheme using the data on the state of the objects reduced to «common time» when the calculations were made.

Existing Case 3 makes it necessary to resolve the following contradiction. Let us examine the three one-type objects O_1, O_2, O_3.

Suppose that

$$\begin{cases} \log x\,(O_1) = \log x\,(O_2) + 1.2 \cdot \delta_{rel}, & O_1 \approx O_2 \\ \log x\,(O_2) = \log x\,(O_3) + 1.2 \cdot \delta_{rel}, & O_2 \approx O_3 \end{cases} \tag{5.9}$$

According to (5.9), $p_{12} = 0.84$ and $p_{23} = 0.84$.

Under the probability theory, the probability p_{13} can never be equal to 1, still the measurements show that

$$\log x\,(O_1) = \log x\,(O_3) + 2.4 \cdot \delta_{rel} > \log x\,(O_3) + 2 \cdot \delta_{rel}, \quad O_1 > O_3. \tag{5.10}$$

It means that for the pair of O_1 and O_3 Case 2 is true, i.e. (1.10) is true, and the object O_1 is undoubtedly more significant than the object O_3. It requires that the whole procedure of estimating objects on the scale of significance be rethought.

In this case one can see a paradox – thanks to an intermediary – the object O_2 the significance of the object O_1, which is higher than the significance of the object O_3, is not absolutely unconditional.

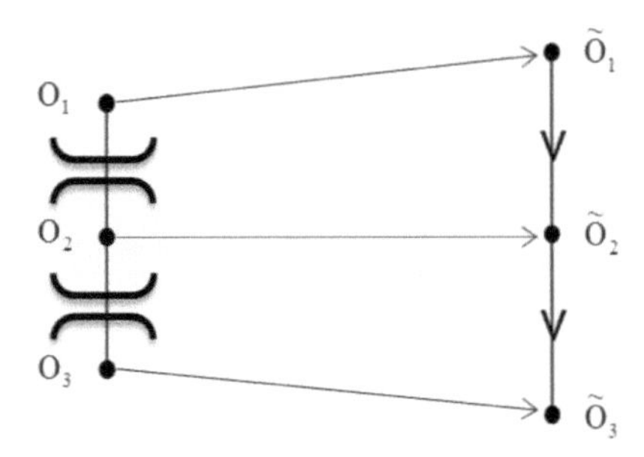

Figure 5.4 The case of equal expert estimates for the two types of objects under comparison.

The position of the objects O_1, O_2 и O_3 on the scale of significance requires a more precise definition, which is possible only by comparing the distinguished objects with some other objects which are likely to be of a different type.

Let us examine the possible cases.

Case A. Each of the objects is expertly recognized as equal to the relevant object $O_1 \sim \widetilde{O}_1$, $O_2 \sim \widetilde{O}_2$, $O_3 \sim \widetilde{O}_3$ (Figure 5.4).

That said, in accordance with the measurement scale (the scheme used to calculate estimates):

$$\begin{cases} \log \widetilde{x}\left(\widetilde{O}_1\right) > \log \widetilde{x}\left(\widetilde{O}_2\right) + 2 \cdot \widetilde{\delta}_{rel} \\ \log \widetilde{x}\left(\widetilde{O}_2\right) > \log \widetilde{x}\left(\widetilde{O}_3\right) + 2 \cdot \widetilde{\delta}_{rel} \end{cases} \tag{5.11}$$

In this case the estimates $x\left(O_1\right)$ and $x\left(O_3\right)$ should be shifted from the estimate $x\left(O_2\right)$, i.e.

$$\begin{cases} \log x\left(O_1\right) = \log x\left(O_2\right) + 2 \cdot \delta_{rel} \\ \log x\left(O_3\right) = \log x\left(O_2\right) - 2 \cdot \delta_{rel} \end{cases} \tag{5.12}$$

The object O_1 is considered to be underestimated, and the object O_3 is considered to be overestimated. The equations $O_1 \sim O_2$ and $O_2 \sim O_3$ are thus replaced with the inequations $O_1 > O_2$ и $O_2 > O_3$. The contradiction ñ $O_1 > O_3$ disappears.

Case B. All the objects are expertly recognized as equal to one and the same object $\widetilde{O}_2$ (Figure 5.5).

It means that the values $x\left(O_1\right)$ and $x\left(O_3\right)$ should be «moved up» to each other and to the estimate $x\left(O_2\right)$.

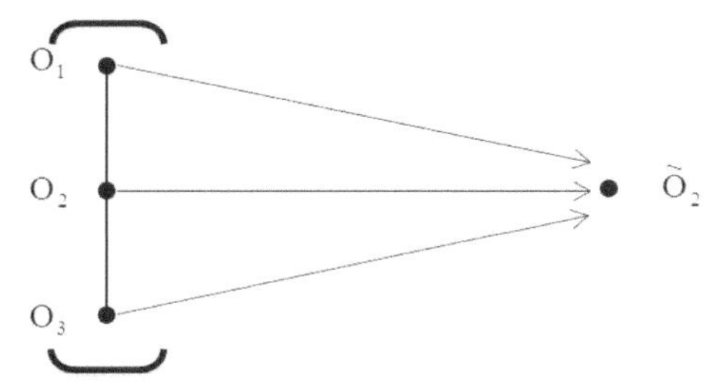

Figure 5.5 The case of the two one-type objects being equal to one object of the second type.

$$\begin{cases} \log x\,(O_1) = \log x\,(O_2) + 1 \cdot \delta_{rel} \\ \log x\,(O_3) = \log x\,(O_2) - 1 \cdot \delta_{rel} \end{cases} \tag{5.13}$$

It will help «eliminate» the contradiction $O_1 > O_3$. The resulting system $O_1 \approx O_2$, $O_2 \approx O_3$, $O_1 \approx O_3$ ceases to be perceived as a contradiction.

Mixed versions of comparison made with the help of another scale are also possible. Let us observe one of them (Figure 5.6).

Case B. The first two objects are expertly recognized as equal to one and the same object $\widetilde{O}_1$. The third object is equal to a less significant $\widetilde{O}_2$. In this case the estimated $\log x\,(O_3)$ and $\log x\,(O_2)$ должны быть раздвинуты. should be widened. Therefore, two domains of attraction are formed.

The analyzed cases (A, B, and C) show that inaccurate estimation that emerges when an inaccurate equation is taken as a criterion for the comparison leads to an urgent grouping (clustering) of the one-type objects O_1,O_2,O_3 that are to be estimated by comparing them with the objects of another type.

It is worth noting that the objects O_1,O_2,O_3 were basic ones, whereas $\widetilde{O}_1$,$\widetilde{O}_2$,$\widetilde{O}_3$ were auxiliary ones. Still, the process is «mirror-like»: as the

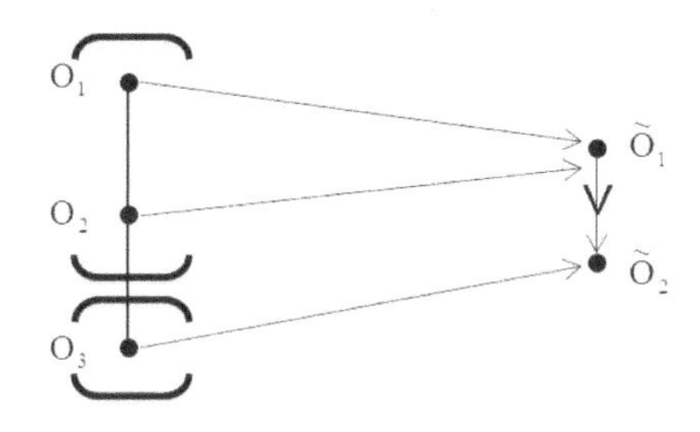

Figure 5.6 A special case of one-type objects being equal to one object of the second type.

estimates of the objects of the first type O_1, O_2, O_3 are specified through the estimates of the objects of the second type $\tilde{O}_1, \tilde{O}_2, \tilde{O}_3$ so as the estimates of the second type $\tilde{O}_1, \tilde{O}_2, \tilde{O}_3$ may be specified through the estimates of the objects of the first type O_1, O_2, O_3.

Moreover, such approach enables the use of more than two scales to draw a comparison. Beyond debate, the use of more than two scales will imply certain peculiarities, e.g. those connected with «the competition» between the second and the third scales while the cluster boundaries of the scale of the first-type objects are being formed. However, the above-stated idea led to the development of an algorithm for an integral estimate of systematic significance of the objects of different types.

Every type of objects that is different from others has its own estimating methods. Needless to say, as a result one gets ranked lists of objects that differ from each other and feature different measures of inaccuracy, and thus, different distinctive cluster sizes.

The normalization of cluster position and the projection of real data (the estimates of objects that are found in clusters) into an interval defined by cluster boundaries allows us to satisfy all the conditions: the compared objects are treated as approximately equal since they are found in one cluster, and sorting objects from the most significant to the least significant is assured by the order of cluster-forming.

The calculated scheme of the below-explained algorithm will present a detailed step-by-step description of the calculation formulae. The cluster composition is given in Table 5.1.

The scheme given in Figure 5.7 provides a general idea of an ultimate solution. It shows an *Y*-axis that contains the estimates of a general integral scale that demonstrates the significance of the objects of three types. Certain graphs meant for the objects of different types have digit indicators being

Table 5.1 The cluster composition

Cluster Name	Cluster Composition
B	z_1
C	$x_1, x_2, x_3; y_1, y_2, y_3, y_4; z_1, z_2$
D	x_4, x_5
E	x_5, z_4
F	x_6
G	$x_7, x_8, x_9; y_6; z_5, z_6, z_7, z_8$
H	x_{10}
I	y_7, y_9

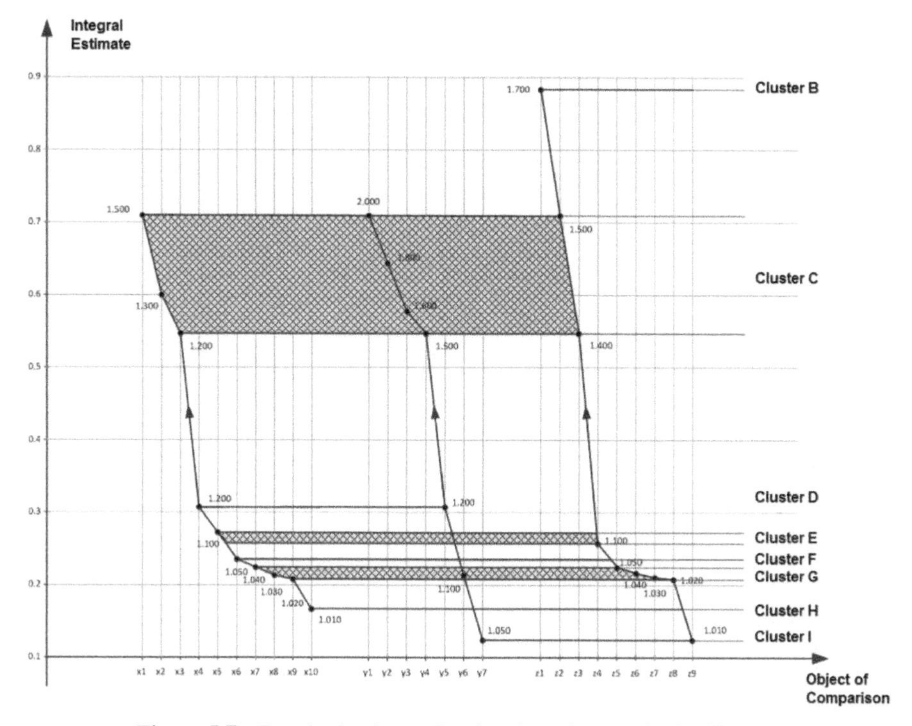

Figure 5.7 Developing integral estimation of systemic significance.

applied – they are inherent background estimates of the significance of the objects. Horizontal lines and Latin letters mark the clusters which combine «approximately equisignificant» objects.

The algorithm itself includes two parts. The first one deals with preparing the background data for calculations. The second one enables the estimation of the significance of the objects in all types of measurement scales at one's disposal, and the ultimate convolution of these estimates into a dimensionless integral estimate of the significance of objects.

5.1.3 Case Study

Let us consider practical use of the algorithm described earlier by a conditional example. This section completely repeats our earlier work [17].

Step 1. Rating of objects in one type is carried out. The example of such rating is presented in Table 5.2.

Table 5.2 Objects estimates in their own importance scale

Object Name	Object Estimate	Object Name	Object Estimate	Object Name	Object Estimate
GCS1	10.96	GDS1	12.59	UGS1	11.22
GCS2	15.85	GDS2	63.10	UGS2	10.23
GCS3	10.23	GDS3	31.62	UGS3	50.12
GCS4	31.62	GDS4	11.22	UGS4	12.59
GCS5	11.22	GDS5	15.85	UGS5	10.96
GCS6	10.47	GDS6	100.00	UGS6	10.47
GCS7	10.72	GDS7	39.81	UGS7	25.12
GCS8	11.75			UGS8	10.72
GCS9	19.95			UGS9	31.62
GCS10	12.59				

For convenience objects types are given the mnemonic name (column 7 of Table 5.3) and two fictitious are added for each type of the value: *TOP* (further-objects) and *BOTTOM* (further *B*-objects) – precalculated borders of possible change of all objects estimates in the corresponding measurement scale. Loaded estimates for each type of data are obtained with own ratio error (column 8) that allows to determine the limit maximum size of a cluster in measurement scales of the corresponding type (column 9).

For calculation of upper and lower bounds of estimation scales knowing maximum and minimum values, we define by means of addition of value of the corresponding cluster to maximum value for T-objects and subtraction of

Table 5.3 Borders of change of system indexes for objects of different types

Type number	Objects type	Objects quantity	Quantity of objects together with boundary objects	Index of the beginning of area of placement of data on objects of the specified type	Index of the end of area of placement of data on objects of the specified type	Mnemonic name of objects type	Measurement error ($\delta_{rel.,m}$)	Size of cluster (Δ_m)	Ponderability of set type scale (ρ_m)
1	2	3	4	5	6	7	8	9	10
1	GCS	10	12	1	12	X	7.9%	0.20	0.600
2	GDS	7	9	13	21	Y	6.3%	0.10	0.300
3	UGS	9	11	22	32	Z	15.8%	0.50	0.100.
Total		26	32	1	32				1.000

Table 5.4 Algorithm output in the early steps

Object Name	Object Estimate	Denary Logarithm of Estimate	Mnemonic Name of Object	Object Index
1	2	3	4	5
T_GCS		1.70	TX	1
GCS1	10.96	1.04	X7	8
GCS2	15.85	1.20	X3	4
GCS3	10.23	1.01	X10	11
GCS4	31.62	1.50	X1	2
GCS5	11.22	1.05	X6	7
GCS6	10.47	1.02	X9	10
GCS7	10.72	1.03	X8	9
GCS8	11.75	1.07	X5	6
GCS9	19.95	1.30	X2	3
GCS10	12.59	1.10	X4	5
B_GCS		0.81	BX	12

the size of the corresponding cluster from minimum value for *B*-objects (for example for objects of the first type $TX = X1 + \Delta_X; BX = X10 + \Delta_m$).

Closeness in estimation of object in this or that scale depends both on the cluster size, and on quantity of objects of that basic type which corresponds to a scale and through which values of estimates, estimates of objects of other types are recalculated. If in any scale the error is big or objects are few, it is obvious that estimates of all objects in this scale will have less "confidence" than more exact estimates for bigger quantity of objects. To consider the specified inadequacy of the used scales, in expert terms the ponderability index ρ_m is attributed to scales and this index reflects an object estimate contribution share in the specified scale into an integrated estimate of system importance.

Step 2. Objects estimates take the logarithm (for example, column 3 Table 5.4 for the first type of objects).

Step 3. Objects of each type in own scales are ordered on decrease of estimates that take the logarithm (column 5, Table 5.4).

Step 4. After all tables' data sorting (one for each objects type) it is possible to unite data with insert *T*-objects and *B*-objects into the united list in the following order:

*TX, X*1, ..., *X*10, *BX*; *TY, Y*1, ..., *Y*10, *BY*; *TZ, Z*1, ..., *Z*10, *BZ.* Number of place thereby becomes the universal system index which definitely specifies both object type and object itself. Creation of general list allows to make calculations with use of these objects without obvious indication of object type. Object index for "subsequent interpretation" includes all necessary information. Such form of data presentation we will call canonical. Values of object in the created united column, we will mark as $\|O_j\|_{[0]}$.

Step 5. Expert comparisons of objects are of different types among themselves. The organization of an appointment procedure of estimate to experts is independent task which isn't considered in this article. Experts aren't obliged to know either the order of types which we chose for expert evaluation, or value of intrasystem indexes. It is desirable that they knew results of objects comparisons given by other experts. Questions of coordination of different expert estimates in this article are not considered. It is supposed that a resultant estimate of pairs – the estimate coordinated by all experts on the results of discussion.

So, for the procedure of "sewing together of lists" initial material is the list of "approximate equalities" (fuzzy equivalences) made by experts (column 2, Table 5.5). Thus it isn't important that any compared objects in this table is absent, and some objects are present several times. As in the considered system there are only "equality" rules, therefore, the summary table a priori doesn't contain internal contradictions. Besides, to each object O the system index is unambiguously attributed, for the subsequent processing to each rule there corresponds record of "expert equality" in the form of the list consisting of three arguments.

For example, the value of the first argument equal to 2 indicates paired objects comparison. Further there are indexes of the objects participating in

Table 5.5 Summary table of expert objects comparison

Rule	Equalities Set by Experts	Record in Mnemonic Names	Record in System Objects Indexes
1	2	3	4
1	GCS2~GDS6	X3=Y1	{2,4,14}
2	GCS10~GDS5	X4=Y5	{2,5,18}
3	GCS6*~GDS1*	X9=Y6	{2,10,19}
4	GCS4~UGS7	X1=Z3	{2,2,25}
5	GCS8~UGS4	X5=Z4	{2,6,26}
6	GCS1~UGS6	X7=Z8	{2,8,30}
7	GCS6*~UGS1*	X9=Z5	{2,10,27}
8	GDS3~UGS9	Y4=Z2	{2,17,24}
9	GDS1*~UGS1*	Y6=Z5	{2,19,27}
10	GDS4~UGS2	Y7=Z9	{2,20,31}

Table 5.6 Canonical form of record of comparisons rules

Rule	Equalities Set by Experts	Record in Mnemonic Names	Record in System Objects Indexes
1	GCS4∼UGS7	X1=Z3	{2,2,25}
2	GCS2∼GDS6	X3=Y1	{2,4,14}
3	GCS10∼GDS5	X4=Y5	{2,5,18}
4	GCS8∼UGS4	X5=Z4	{2,6,26}
5	GCS1∼UGS6	X7=Z8	{2,8,30}
6	GCS6∼GDS1∼UGS1	X9=Y6=Z5	{3,10,19,27}
7	GDS3∼UGS9	Y4=Z2	{2,17,24}
8	GDS4∼UGS2	Y7=Z9	{2,20,31}

equality. So, equality 8, for example, contains the mentioned GDS3 (*Y*4, system index 17). The expert specified that it is approximately equal in importance to UGS9 (*Z*2, system index 24).

Step 6. As in the compared pairs there can be objects which were exposed to comparison more than once ("star"), for the purpose of optimization of the subsequent calculations on this step all compared to the same object (for example with *X*9 (KC6)) are united into one list (in Table 5.6, line 6).

As a result, record in system indexes take the form {3,10,19,27} that is equivalent to the record GCS6∼GDS1∼UGS1.

Such form of record is more economic, as with growth of the size of a chain of equivalent objects (*L*) one record $\{L, \arg 1, \ldots, \arg L\}$ replaces $\frac{L(L-1)}{2}$ of records of paired comparisons.

Step 7. Calculation of estimates of the integrated importance of objects assumes calculation of estimates of each object in all types of scales. To receive the estimate in any scale it is necessary to estimate borders of possible changes of values of these estimates, that is to obtain: the lower guaranteed estimates for maximum value of estimates which objects – $||\mathrm{O}_j||_{[m]}^{\max}$ can have, and the upper guaranteed estimates for minimum value of estimates which the same objects $||\mathrm{O}_j||_{[m]}^{\min}$ can have. The group of objects (which can consist even of one object) in which values of pairs $\left\{||\mathrm{O}_j||_{[m]}^{\max}, ||\mathrm{O}_j||_{[m]}^{\min}\right\}$ coincide are in fact equivalent objects which according to measurement logic must get into one cluster. For effective calculation of the sizes $||\mathrm{O}_j||_{[m]}^{\max}$ and $||\mathrm{O}_j||_{[m]}^{\min}$ it is expedient to create from canonical forms of equalities lists and own canonical forms of estimates 2*M* of working tables of equalities and 2*M* of initial starting values of objects estimates for calculation algorithms of values of the upper $||\mathrm{O}_j||_{[m][start]}^{\max}$ and the lower $||\mathrm{O}_j||_{[m][start]}^{\min}$ bounds. The upper estimates of value of each object O in any scale *m* (*X, Y* or *Z*) $\|O_j\|_{[m]}^{\max}$ are calculated by means of the current objects estimates values raising. For objects of basic type (what type of objects will be considered as a basic one is decided at step 1) coinciding with scale type *m*, starting values are taken equal to values from canonical form.

In Table 5.7 they are marked in bold type and are situated in column 4, 6 and 8 consequently for scale type *X, Y* and *Z*.

To *B*-objects types which aren't coinciding with basic type the values of *T*-object scale type are attributed, i.e. starting *BY* values for 21 objects (*B*-object type *Y*) and *BZ* value for 32 objects (*B*-object of type *Z*) take *BX* values (in the concerned example it is equal to 0.81). All-objects (the 13th object *TY* and the 22nd object *TZ*) take the values equal to *TX* (respectively, 1.70).

Table 5.7 Samples of filling of estimates calculation effective range in every scale types

Field index	Mnemonic name	For calculation clusters lower borders in scale X	For calculation clusters upper borders in scale X	For calculation clusters lower borders in scale Y	For calculation clusters upper borders in scale Y	For calculation clusters lower borders in scale Z	For calculation clusters upper borders in scale Z
1	2	3	4	5	6	7	8
1	TX	1.700	1.700	TY	TY	TZ	TZ
2	X1	1.500	1.500	TY	BY	TZ	BZ
3	X2	1.300	1.300	TY	BY	TZ	BZ
4	X3	1.200	1.200	TY	BY	TZ	BZ
5	X4	1.100	1.100	TY	BY	TZ	BZ
6	X5	1.070	1.070	TY	BY	TZ	BZ
7	X6	1.050	1.050	TY	BY	TZ	BZ
8	X7	1.040	1.040	TY	BY	TZ	BZ
9	X8	1.030	1.030	TY	BY	TZ	BZ
10	X9	1.020	1.020	TY	BY	TZ	BZ
11	X10	1.010	1.010	TY	BY	TZ	BZ
12	BX	0.810	0.810	BY	BY	BZ	BZ
13	TY	TX	TX	2.100	2.100	TZ	TZ
14	Y1	TX	BX	2.000	2.000	TZ	BZ
15	Y2	TX	BX	1.800	1.800	TZ	BZ
16	Y3	TX	BX	1.600	1.600	TZ	BZ
17	Y4	TX	BX	1.500	1.500	TZ	BZ
18	Y5	TX	BX	1.200	1.200	TZ	BZ
19	Y6	TX	BX	1.100	1.100	TZ	BZ
20	Y7	TX	BX	1.050	1.050	TZ	BZ
21	BY	BX	BX	0.950	0.950	BZ	BZ

22	TZ	TX	TX	TY	TY	2.200	2.200
23	Z1	TX	BX	TY	BY	1.700	1.700
24	Z2	TX	BX	TY	BY	1.500	1.500
25	Z3	TX	BX	TY	BY	1.400	1.400
26	Z4	TX	BX	TY	BY	1.100	1.100
27	Z5	TX	BX	TY	BY	1.050	1.050
28	Z6	TX	BX	TY	BY	1.040	1.040
29	Z7	TX	BX	TY	BY	1.030	1.030
30	Z8	TX	BX	TY	BY	1.020	1.020
31	Z9	TX	BX	TY	BY	1.010	1.010
32	BZ	BX	BX	BY	BY	0.510	0.510

All other objects of the types which aren't coinciding with basic take *BX* values, as before the algorithm work we can guarantee only the most minimum values of estimates. Please note that if any objects of the types which aren't coinciding with basic during algorithm won't be compared to other objects, their values will remain minimum in this scale of *m* though in their own scale they can be not the most little significant objects.

Similarly, in columns 3, 5, 7 of Table 5.7 there are starting values for calculation of the lower estimates of each object – $\|O_j\|^{\min}_{[m][start]}$. Objects data of basic type are also copied from canonical form.

B-objects and *T*-objects of the types which aren't coinciding with basic one accept values of *B*-object and *T*-object of basic type, and to the rest objects of the types which aren't coinciding with basic type values of *T*-object of basic type are given. Such distinction in marking is caused by that estimates of the lower bounds of objects are calculated by decrease of the current estimates values.

In order that procedures of definition of upper and lower estimates of objects are carried out the most quickly it is necessary to modify comparison rules of canonical table into 2*M* of working tables of comparison rules. Working tables don't differ from canonical table according to the content. It is only necessary to change the order of consideration of comparison rules in them, that is to carry out permutation of lines. The content of six working tables is given in summary Table 5.8.

Step 8. The integral estimate of object importance O – $\|O\|_{int}(O)$ – is calculated through contribution of object estimates O in scales of each of types of objects – $\|O\|_{[m]}$ by formula:

$$\|O\|_{[m]} = \sum_{m} \rho_m \cdot \frac{\|O_j\|_{[m]} - \|B_m\|_{[m]}}{\|T_m\|_{[m]} - \|B_m\|_{[m]}} \qquad (*)$$

Table 5.8 Working tables of rules record for procedures of calculation of objects clusters borders in all types of scales

Rule	Record in mnemonic names	Working table for definition of upper estimates of objects in scale X [4]	Rule	Record in mnemonic names	Working table for definition of lower estimates of objects in scale X [3]
1	X1=Z3	{2,2,25}	1	X9=Y6=Z5	{3,10,19,27}
2	X3=Y1	{2,4,14}	2	X7=Z8	{2,8, 30}
3	X4=Y5	{2,5,18}	3	X5=Z4	{2,6,26}
4	X5=Z4	{2,6,26}	4	X4=Y5	{2,5,18}
5	X7=Z8	{2,8,30}	5	X3=Y1	{2,4,14}
6	X9=Y6=Z5	{3,10,19,27}	6	X1=Z3	{2,2, 5}
7	Y4=Z2	{2,17,24}	7	Y7=Z9	{2,20,31}
8	Y7=Z9	{2,20,31}	8	Y4=Z2	{2,17,24}
1	Y1=X3	{2,14,4}	1	Y7=Z9	{2,20,31}
2	Y4=Z2	{2,17,24}	2	Y6=X9=Z5	{3,19,10,27}
3	Y5=X4	{2,18,5}	3	Y5=X4	{2,18,5}
4	Y6=X9=Z5	{3,19,10,27}	4	Y4=Z2	{2,17,24}
5	Y7=Z9	{2,20,31}	5	Y1=X3	{2,14,4}
6	X1=Z3	{2,2,25}	6	X7=Z8	{2,8,30}
7	X5=Z4	{2,6,26}	7	X5=Z4	{2,6,26}
8	X7=Z8	{2,8,30}	8	X1=Z3	{2,2,25}
1	Z2=Y4	{2,24,17}	1	Z9=Y7	{2,31,20}
2	Z3=X1	{2,25,2}	2	Z8=X7	{2,30,8}
3	Z4=X5	{2,26,6}	3	Z5=X9=Y6	{3,27,10,19}
4	Z5=X9=Y6	{3,27,10,19}	4	Z4=X5	{2,26,6}
5	Z8=X7	{2,30,8}	5	Z3=X1	{2,25,2}
6	Z9=Y7	{2,31,20}	6	Z2=Y4	{2,24,17}
7	X3=Y1	{2,4,14}	7	X4=Y5	{2,5,18}
8	X4=Y5	{2,5,18}	8	X3=Y1	{2,4,14}

where $\|B_m\|_{[m]}$ – value of BOTTOM-object in scale type *m*; $\|T_m\|_{[m]}$ – value of TOP-object in scale type *m*; ρ_m – a control indicator of ponderability of scale type *m* (Table 5.1 column (x)); $\|O_j\|_{[m]}$ – O object estimate in scale type *m*.

By formula (*) for all types of scales (in our case for scales of objects of type *X*, *Y* and *Z*) to execute a number of additional steps for receiving estimates $\|O\|_{[m]}$. We will illustrate the sequence of actions by the example of calculations of O objects estimates in scale type *X*.

To receive the required result given in column 4 Table 5.9 it is necessary to consider the following properties of estimate: estimate value of any object must be increased on two bases:

Table 5.9 Definition of upper estimates of objects in scale X

	Scale Type X	Cyclic implementation of comparison rules with alignment of arguments to the maximum values	Upper estimate of objects in scale X
1	TX	1.70	1.70
2	X1	1.50	1.50
3	X2	1.30>1.50(2/2 cycle)	1.50
4	X3	1.20>1.50(2/2 cycle)	1.50
5	X4	1.10	1.10
6	X5	1.07	1.07
7	X6	1.05	1.05
8	X7	1.04	1.04
9	X8	1.03>1.04(6/1 cycle)	1.04
10	X9	1.02>1.04(6/1 cycle)	1.04
11	X10	1.01	1.01
12	BX	0.81	0.81
13	TY	1.70	1.70
14	Y1	0.81>1.20(2/1 cycle)>1.50(7/1 cycle)	1.50
15	Y2	0.81>1.10(3/1 cycle) >1.50(7/1 cycle)	1.50
16	Y3	0.81>1.10(3/1 cycle) >1.50(7/1 cycle)	1.50
17	Y4	0.81>1.10(3/1 cycle)>1.50(7/1 cycle)	1.50
18	Y5	0.81>1.10(3/1 cycle)	1.10
19	Y6	0.81>1.04(6/1 cycle)	1.04
20	Y7	0.81	0.81
21	BY	0.81	0.81
22	TZ	1.70	1.70
23	Z1	0.81>1.50(1/1 cycle)	1.50
24	Z2	0.81>1.50(1/1 cycle)	1.50
25	Z3	0.81>1.50(1/1 cycle)	1.50
26	Z4	0.81>1.07(4/1 cycle)	1.07
27	Z5	0.81>1.04(5/1 cycle)	1.04
28	Z6	0.81>1.04(5/1 cycle)	1.04
29	Z7	0.81>1.04(5/1 cycle)	1.04
30	Z8	0.81>1.04(5/1 cycle)	1.04
31	Z9	0.81	0.81
32	BZ	0.81	0.81

- object, owing to comparison rules gets value bigger than current, from other object with which it is compared in comparison rules;
- object, more significant in its own scale, in any of scales can't have value less than less significant object in its own scale.

Step 8.1. Consistently, from the first to the last rule in the working table it is necessary to carry out check of arguments on equality of estimates values. In case of non-performance of equality in the rule correction of smaller values of arguments to the maximum value is made.

Table 5.10 Calculation of integral estimate of objects through their estimates in all private scales

Field Index	Mnemonic Name	Objects Estimates in Their Own Scales	Object Estimate in Scale X	Object Estimate in Scale Y	Object Estimate in Scale Z	Integral Estimate of Object
1	2	3	4	5	6	7
1	TX	1.700	1.700	2.100	2.200	1.000
2	X1	1.500	1.450	1.800	1.500	0.711
3	X2	1.300	1.317	1.733	1.433	0.600
4	X3	1.200	1.250	1.700	1.400	0.546
5	X4	1.100	1.100	1.200	1.250	0.304
6	X5	1.070	1.070	1.200	1.100	0.275
7	X6	1.050	1.050	1.100	1.075	0.234
8	X7	1.040	1.040	1.100	1.050	0.226
9	X8	1.030	1.030	1.100	1.035	0.218
10	X9	1.020	1.020	1.100	1.020	0.211
11	X10	1.010	1.010	1.025	0.765	0.170
12	BX	0.810	0.810	0.950	0.510	0.000
13	TY	2.100	1.700	2.100	2.200	1.000
14	Y1	2.000	1.450	1.800	1.500	0.711
15	Y2	1.800	1.370	1.760	1.460	0.644
16	Y3	1.600	1.290	1.720	1.420	0.579
17	Y4	1.500	1.250	1.700	1.400	0.546
18	Y5	1.200	1.100	1.200	1.250	0.304
19	Y6	1.100	1.030	1.100	1.035	0.218
20	Y7	1.050	0.915	1.050	1.010	0.127
21	BY	0.950	0.810	0.950	0.510	0.000
22	TZ	2.200	1.700	2.100	2.200	1.000
23	Z1	1.700	1.600	2.050	1.700	0.889
24	Z2	1.500	1.450	1.800	1.500	0.711
25	Z3	1.400	1.250	1.700	1.400	0.546
26	Z4	1.100	1.070	1.150	1.100	0.262
27	Z5	1.050	1.040	1.100	1.050	0.227
28	Z6	1.040	1.033	1.100	1.040	0.221
29	Z7	1.030	1.027	1.100	1.030	0.216
30	Z8	1.020	1.020	1.100	1.020	0.211
31	Z9	1.010	0.915	1.050	1.010	0.127
32	BZ	0.510	0.810	0.950	0.510	0.000

Step 8.2. To carry out check of need of change of objects estimates values on the second basis, but only for those objects which stand directly above the object modified on the first basis. If during check of all working list of rules at least one adjustment was carried out (on the end of cycle), the cycle of rules check should be renewed.

Table 5.11 Rating objects list

Rating	Object Name	Integral Estimate of Object	Rating	Object Name	Integral Estimate of Object
1	Z1	0.889	10	X6	0.234
2	X1, Y1, Z2	0.711	11	Z5	0.227
3	Y2	0.644	12	X7	0.226
4	X2	0.600	13	Z6	0.221
5	Y3	0.579	14	X8, Y6	0.218
6	X3, Y4, Z3	0.546	15	Z7	0.216
7	X4, Y5	0.304	16	X9, Z8	0.211
8	X5	0.275	17	X10	0.170
9	Z4	0.262	18	Y7, Z9	0.127

Step 8.3. All list of objects divided into subgroups which we form by sorting pair <the upper estimate of object, the lower estimate of object> on decrease of values of the first argument, and at equality of the first arguments, in addition on decrease of the second argument of the specified pair. In that case, when the difference between estimates of upper and lower bounds of clusters is positive and exceeds the cluster size for the objects estimated in scale of this type it is necessary to analyze clusters at a size of admissible sizes by introduction of correction statement that estimates of the most significant objects of a cluster are overestimated, and estimates of the least significant objects of a cluster are underestimated. If in this type of objects there is a unique object entering a cluster, it must be always placed in the middle of a cluster.

Step 9. The integrated estimate of object importance (column 7 Table 5.10) is calculated by a formula (*) on the basis of estimates in private scales of *X*, *Y* and *Z* (column 4, 5 and 6).

Final rating objects list is given in Table 5.11.

5.2 The Analytic Hierarchy Process Modification for Decision Making under Uncertainty

5.2.1 Introduction

This section completely repeats our earlier work [18]. Huge number of tasks in different fields of knowledge (such as: definition of relative weight purposes during carrying out multicriteria optimization; definition of expected result of activity, for example determination of investment project efficiency; definition of probabilities of implementation of accident development various scenarios in potentially dangerous object in the probabilistic analysis of its safety; a task of probabilities of outcomes in matrix of consequences at a choice of rational decision in the conditions of partial uncertainty, etc.) are solved in conditions when the behavior of studied process branches and it is necessary to define weights (relative probabilities) of possible scenarios (compared objects). In other tasks some parameters of mathematical models can accept various values depending on some non-formalized (not described mathematically) factors. Definition of probabilistic distributions of these parameters is often impossible due to the lack of rather representative statistics. For value assignment of these parameters, qualitative or qualitative and quantitative scales are used.

In all considered cases (both a scenario choice, and class definition or object state in some scale) mathematically the task is reduced to compared objects' choice from a set that is possible by means of expert estimation. Whereby the interest is not only in choosing of the most probable compared objects, but also in defining compared objects' weight (relative probabilities). This information is used further in algorithms of simulation modeling, and in simpler models in averaging operations taken into account compared objects' weight.

Simplicity of relative compared objects' weight calculation at their determination accuracy preservation often plays considerable part. The analysis shows that the most reliable and widely used method of problem solution of choice and definition of relative compared objects' weight (relative probabilities) is the pairwise comparisons method on qualitative attribute with quantitative assessment of preferences [19].

5.2.2 Literature Review

The pairwise comparisons method [19] had broad application when determining relative indicators of compared objects' importance against chosen qualitative criterion in decision-making support systems. Concerning some general property for compared objects [20–23] this method allows to carry out their ranging, to define a priority of one compared object (criterion, purpose, alternative) before another. One of the most developed and widely practiced is modification of pairwise comparisons method, received the name of Saaty's Analytic Hierarchy Process (AHP) [24].

One of the essential shortcomings of pairwise comparisons method's various realization is the essential increase in labor input of necessary calculations, with growth of number of estimated objects. In some cases, it is too difficult for an expert to perform a large number of pairwise comparisons or he can't simply compare two objects on the offered preferences scale.

There is a problem of processing not completely certain pairwise comparisons' matrixes. The solution of this task is connected to the following problems [25]. The first problem is in definition of necessary and sufficient conditions of calculation of all components of compared objects' importance indicators vector, and the second one is in definition of the algorithm of these components calculation itself.

In Ref. [26] these problems are solved concerning of pairwise comparisons' results processing in fundamental scale by Eigenvector Following method. The method offered in Ref. [26], in point of fact, is reduced to restoration of full matrix of comparisons on available incomplete matrix and application standard algorithm of relative compared objects' weight to the restored matrix. For method justification it is offered to use graph interpretation of pairwise comparisons method. Tops of graph G designate alternatives $A_i \in A$, $i = \overline{1,k}$ from which arcs d_{ij} release, connected them with each other. It is proved that d_{ij} is equal to geometrical average intensity of all possible ways connecting tops A_i and A_j. Necessary and sufficient condition for this method realization: between any pair of tops in graph G there has to be at least one way.

5.2.3 Problem Statement

We consider the problem of compared objects' weight definition importance according to incomplete pairwise comparisons' matrixes. In practice most precisely pairwise comparisons are given by experts between "the most significant objects and other objects" and between "the least significant objects and others", that is when differences between pairwise compared objects are contrast. When compared objects are close "naturally" to comparisons matrix element, the expert according to the scale [24], assigns value 1 that leads, in our opinion, to unjustified alignment of compared objects' weight estimates.

We offer to consider, in this case, pairwise comparisons uncertain. Whereby instead of the procedure of recovering such uncertain data we offer reverse procedure lying in the fact that to find such subset of pairwise comparisons which is most coordinated and this subset, finally, assigns values of compared objects' weight. Thereby two aims are achieved. First, the required decision in the form of distribution of compared objects' weight importance will be with some set accuracy is coordinated with pairwise comparisons' matrix constructed on AHP method. Secondly, the remained pairwise comparisons will indicate communication circuits between compared objects, explaining distinctions of defined weights.

So, the task is set as follows: not completely-filled pairwise comparisons' matrix is set and some initial (for example, uniform) distribution of weights importance objects W_i is set. The integrated majorant criterion of these weights mismatch and available at every moment submatrix of comparisons matrix is formulated – P.

It is necessary to construct iterative algorithm of ejection of the smallest number of least coordinated pairwise comparisons with simultaneous correction of objects weights W_i so that mismatch criterion P decreases to some acceptable level (probably equal to zero). Final distribution of compared objects' weights, received as a result of such adaptive iterative procedure has to be considered as the required solution.

5.2.4 Methodology Description

Before describing offered algorithm, we will give some important definitions.

Definition 1. N – dimension of problem, amount of compared objects $O_i = \{O_1, \ldots, O_N\}$.

Definition 2. $V = \{v_1, \ldots, v_N\}$ – set of tops of estimated compared objects' communication graph.

Definition 3. $W_i = W(V_i) > 0$ – normalized positive estimate of importance *i* object (compared objects).

Normalization condition:

$$\sum_i W_i = 1. \tag{5.14}$$

Definition 4. $H = H(V, V)$ – $N \times N$ dimensionality communication graph.

Whereby $H_{ij} = 1$, if V_i and V_j are connected, $H_{ij} = 0$, if such connection lacks due to lack of pairwise comparisons of O_i and O_j. It is estimated that communication graph H is related graph.

Definition 5. $G = G(V, V)$ – subgraph, is named spanning tree of graph H, if there is only one way from any top V_k to any top V_l.

Naturally quantity of spanning tree edges (quantity of pairwise comparisons) in spanning tree is equal to $N - 1$. In general spanning tree is given by "snowflake" (Figure 5.8).

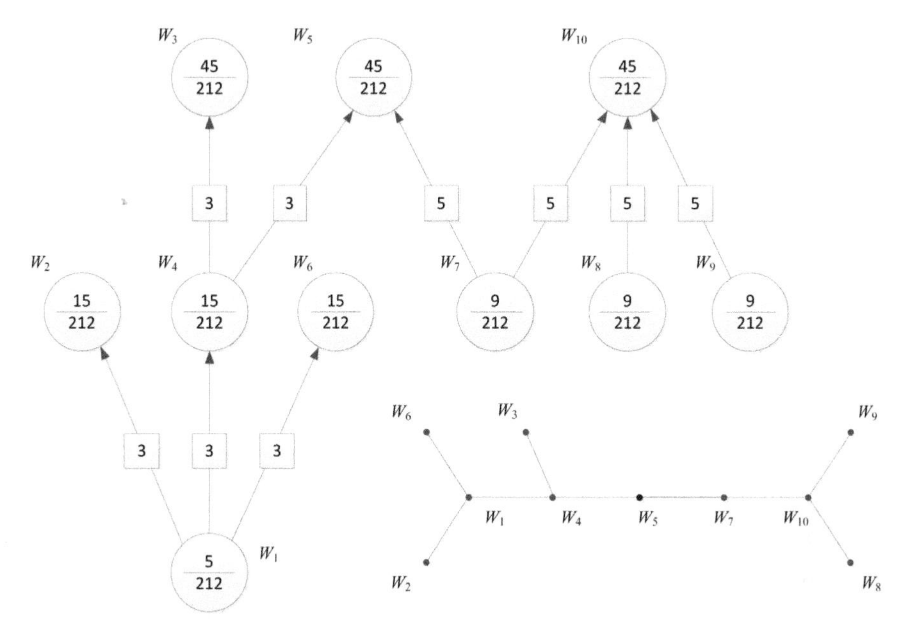

Figure 5.8 Spanning tree (in circles – calculated weights of objects importance (tops), in squares – coefficients of pairwise comparisons of conditional example).

Definition 6. To every non-zero communication graph edge G is given estimate S_{ij}, showing pairwise comparisons

$$W_i = S_{ij} \cdot W_j. \tag{5.15}$$

Please note if spanning tree is chosen then $N - 1$ equality (5.15) and normalization requirement (5.14) uniquely allow to calculate all W_i. According to found W_i estimations $\widetilde{S}_{ij}$ are defined for other communication graph edges not including into spanning tree. The problem is that these estimations are not the same as pairwise comparisons estimations S_{ij}, given by experts.

If connectivity graph is fully connected graph with N tops, relevant to Saati's method then quantity of spanning tree is estimated by Cayley formula with value N^{N-2} [27]. Thereby search for optimal spanning tree in some multitude similarity metrics $\{\widetilde{S}_{ij}\}$ and $\{S_{ij}\}$ by enumeration presents certain technical problems.

Simultaneously it must be admitted that pairwise comparisons estimations given by experts are rather crude and for high dimensions N «mechanistic». Thereby it is incorrectly to require proximate satisfaction of equalities (5.15) for all H graph edges, and in this way for edges entered into spanning tree G.

In this context, we take compelled step – we weaken conditions (5.15), allowing for tops V_i and V_j through S_{ij} mensurate opponent with some default level. Herewith of course it needs to consider that similarly measuring is asymmetric, i.e. the fact is not so critical that the object with big importance gets through another less important object's importance estimation which is higher as it is. Also it is not critical when the object with little importance is underestimated because of more important objects.

Thus, for every communication graph edge we shall estimate value

$$f_{ij} = \frac{W_i}{S_{ij} \cdot W_j} - 1. \tag{5.16}$$

"Ideally" all f_{ij} must be equal to zero, as for us we require them to be in some window with width $2P$. Here P is important parameter of suggested method which while solution construct shall go to zero (note that zero is realizable if we take decision on solution determination specifically in the shape of spanning tree). In addition, not to lose integrating characteristics of pairwise comparisons, it is sufficient to concentrate on some level P_f, different from zero a little. In this situation connected subgraph $H(p)$ of graph H will be formed, exhibiting the characteristic when all estimations f_{ij} are set in window with width $2P_j$.

Taken into account the skewness of estimating f_{ij} for different values we consider two intervals $R(W_i)$ and $L(W_i)$ add up to size P (Figure 5.9):

$$L(W_i) + R(W_i) = 2P. \tag{5.17}$$

We will consider that for big W_i values the center of window displaces to the right of zero and for little values W_i, respectively – to the left of zero. Herewith in "middle" position ($W_i = \frac{1}{N}$) displacement is not occurred. Value of right border $R(W_i)$ under $W_i = 1$ we'll consider equal to $1.9P$, left one, respectively $-0.1P$. To the contrary for little W_i values right border is "forced against" level $0.1P$, and left border stands off zero at a distance of $1.9P$.

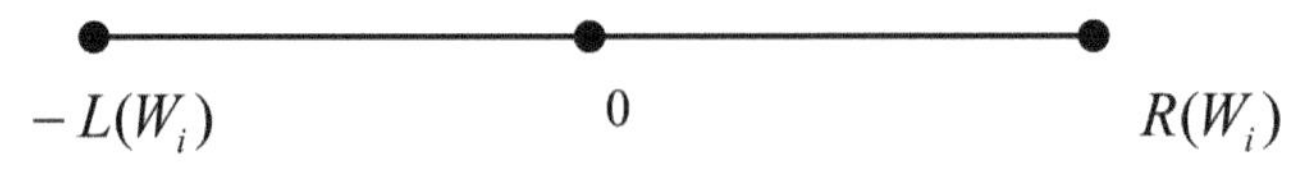

Figure 5.9 The window of set estimations.

We suggest plotting correspondence in the shape of hyperbola:

$$R(W_i) = 0.1P + 1.8P \cdot \frac{(N-1) \cdot W_i}{1 + (N-2) \cdot W_i}. \quad (5.18)$$

At first for big values P in equations

$$-L(W_i) \leq f_{ij} \leq R(W_i) \quad (5.19)$$

are performed for all communication graph edges.

While reduction P exit vertex of one or several estimations f_{ij} will occur to the borders of windows.

While f_{ij} exits to the right border of the window $R(W_i)$ value W_i must be extended, herewith the center of this window will move to the right as simultaneous reduction of all other weights. While f_{ij} exits to the left border, W_i must be reduced with simultaneous increase of weights of all other tops.

Adaptation of tops' weights in G coupling matrix happens. At the cost of this adaptation further reduction of P value contributes, which as the final result ends in situation when all tops W_i have even one edge situated according to its estimation f_{ij} at the border of relevant window. Correlation conflict happens. It is impossible to move even one edge G_{ij} inside window not "pushing out" any other edge G_{kl} in the amount of reached default level P.

In this situation one of the critical edges must be moved off coupling matrix. It is clear this edge mustn't be spanning, i.e. G graph connectivity mustn't be lost.

By experiment we move off such edge G_{ij}, for which another edge G_{ik} exists very close to the border of the window.

As a result all f_{ij} adjoined to top V_i constitute content to relevant border of the window, and make it possible to continue adaptation process, changing both value W_i and value W_j, by virtue of the fact that V_j after critical coupling release also became "less bounded".

Omitting detailed descriptions, we give the general view scheme of method under discussion (Figure 5.10).

5.2.5 Case Study

We will consider control-flow chart operation by the example of pairwise comparisons' matrix processing of ten objects. Basic data are given in Table 5.12.

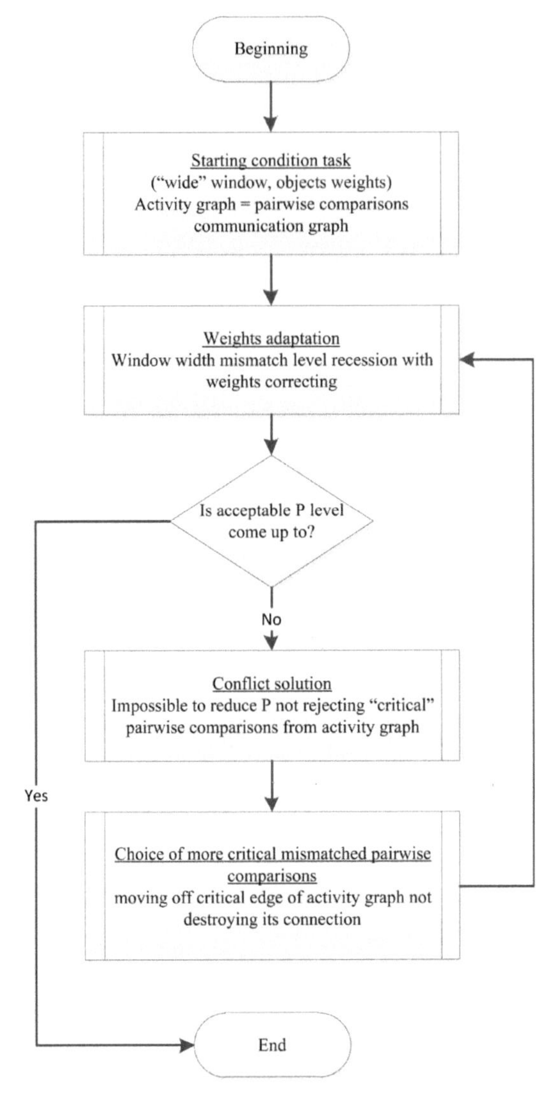

Figure 5.10 Book keeping scheme algorithm.

From 45 pairwise comparisons 18 comparisons are passed (NA). Other 27 estimations indicate preference of objects O3 and O5 over other objects, but mismatch of these estimations demanded clarification of the following circumstance: whether these estimations are enough to exclude demonstratively from leaders, let us say, object O10. We will note that the example is

Table 5.12 Example of pairwise comparisons' matrix processing of ten objects

	O1	O2	O3	O4	O5	O6	O7	O8	O9	O10
O1		$\frac{1}{3}$	$\frac{1}{5}$	$\frac{1}{3}$	$\frac{1}{7}$	$\frac{1}{3}$	NA	NA	NA	$\frac{1}{4}$
O2	$\frac{3}{1}$		$\frac{1}{5}$	NA	$\frac{1}{5}$	$\frac{1}{4}$	NA	NA	NA	$\frac{1}{3}$
O3	$\frac{5}{1}$	$\frac{5}{1}$		$\frac{3}{1}$	NA	$\frac{3}{1}$	$\frac{3}{1}$	$\frac{5}{1}$	$\frac{5}{1}$	$\frac{3}{1}$
O4	$\frac{3}{1}$	NA	$\frac{1}{3}$		$\frac{1}{3}$	NA	NA	NA	NA	$\frac{1}{3}$
O5	$\frac{7}{1}$	$\frac{5}{1}$	NA	$\frac{3}{1}$		$\frac{3}{1}$	$\frac{5}{1}$	$\frac{5}{1}$	$\frac{5}{1}$	$\frac{3}{1}$
O6	$\frac{3}{1}$	$\frac{4}{1}$	$\frac{1}{3}$	NA	$\frac{1}{3}$		NA	NA	NA	$\frac{1}{4}$
O7	NA	NA	$\frac{1}{3}$	NA	$\frac{1}{5}$	NA		NA	NA	$\frac{1}{5}$
O8	NA	NA	$\frac{1}{5}$	NA	$\frac{1}{5}$	NA	NA		NA	$\frac{1}{5}$
O9	NA	NA	$\frac{1}{5}$	NA	$\frac{1}{5}$	NA	NA	NA		$\frac{1}{5}$
O10	$\frac{4}{1}$	$\frac{3}{1}$	$\frac{1}{3}$	$\frac{3}{1}$	$\frac{1}{3}$	$\frac{4}{1}$	$\frac{5}{1}$	$\frac{5}{1}$	$\frac{5}{1}$	

taken from authors' real task decision practice, instead of specially prepared for illustration of offered method opportunities.

So, initially choosing weights of all objects equal 0,1 through an exception of edges with simultaneous reduction of the size of the window P, the decision $W(O)$ provided in Figure 5.8 and in two last columns of Table 5.13 (in the form of standard fraction and in a decimal form) is constructed.

In Table 5.13 we can find the summary data illustrating that estimations mismatch on different steps of algorithm differently influences compared objects' importance weights ratios. So, in process of critical edges removal defining these edges estimations mismatch with all set of pairwise comparisons which have remained in working subgraph, these weights' values execute a periodic motion relating to final solution given in the last column.

It should be noted that coordination of estimates of objects at $P = 0.742$ level (the second column of Table 5.13) and at $P = 0.469$ level (the third column of Table 5.13) demanded removal of mismatched edges, however begin with $P = 0.207$ level all mismatches were found eliminated and further weights corrections were carried out only at the cost of "revaluation of objects importance weights" adaptation mechanism.

It is also important to note that the first mismatches conflict occurred at $P = 1.15$ level. To overcome this conflict communication estimation O3–O10 was removed. Thus succeedent number of removals occurs as if the algorithm in advance knows the solution constructed on spanning tree edges.

Table 5.13 The summary of data illustrating estimations mismatch on different steps of algorithm

	Window size							
Weight	$P =$ 0,742	$P =$ 0,469	$P =$ 0,207	$P =$ 0,1	$P =$ 0,05	$P =$ 0,01	$P =$ 0,0001	$P = 0$
W1	0,047	0,0447	0,026	0,026	0,025	0,0237	0,0235	$\frac{5}{212}$ = 0,02359
W2	0,084	0,0923	0,077	0,075	0,073	0,0711	0,0707	$\frac{15}{212}$ = 0,07075
W3	0,209	0,1735	0,185	0,210	0,211	0,2110	0,2123	$\frac{45}{212}$ = 0,21226
W4	0,084	0,0923	0,077	0,075	0,073	0,0711	0,0707	$\frac{15}{212}$ = 0,07075
W5	0,209	0,1735	0,185	0,210	0,211	0,2110	0,2123	$\frac{45}{212}$ = 0,21226
W6	0,084	0,0923	0,077	0,075	0,073	0,0711	0,0707	$\frac{15}{212}$ = 0,07075
W7	0,047	0,0357	0,045	0,042	0,042	0,0425	0,0425	$\frac{9}{212}$ = 0,04246
W8	0,031	0,0357	0,045	0,042	0,042	0,0425	0,0425	$\frac{9}{212}$ = 0,04246
W9	0,031	0,0357	0,045	0,042	0,042	0,0425	0,0425	$\frac{9}{212}$ = 0,04246
W10	0,174	0,2243	0,238	0,203	0,208	0,2135	0,2123	$\frac{45}{212}$ = 0,21226

Table 5.14 Illustration of algorithm action

	O1	O2	O3	O4	O5	O6	O7	O8	O9	O10
O1		$\frac{0}{1260}$	$-\frac{560}{1260}$	$\frac{0}{1260}$	$\frac{280}{1260}$	$\frac{0}{1260}$	NA	NA	NA	$-\frac{700}{1260}$
O2	$\frac{0}{1260}$		$+\frac{840}{1260}$	NA	$+\frac{840}{1260}$	$+\frac{3780}{1260}$	NA	NA	NA	$\frac{0}{1260}$
O3	$+\frac{1008}{1260}$	$-\frac{504}{1260}$		$\frac{0}{1260}$	NA	$\frac{0}{1260}$	$+\frac{840}{1260}$	$\frac{0}{1260}$	$\frac{0}{1260}$	$-\frac{840}{1260}$
O4	$\frac{0}{1260}$	NA	$\frac{0}{1260}$		$\frac{0}{1260}$	NA	NA	NA	NA	$\frac{0}{1260}$
O5	$+\frac{360}{1260}$	$-\frac{504}{1260}$	NA	$\frac{0}{1260}$		$\frac{0}{1260}$	$\frac{0}{1260}$	$\frac{0}{1260}$	$\frac{0}{1260}$	$-\frac{840}{1260}$
O6	$\frac{0}{1260}$	$-\frac{945}{1260}$	$\frac{0}{1260}$	NA	$\frac{0}{1260}$		NA	NA	NA	$+\frac{420}{1260}$
O7	NA	NA	$-\frac{504}{1260}$	NA	$\frac{0}{1260}$	NA		NA	NA	$\frac{0}{1260}$
O8	NA	NA	$\frac{0}{1260}$	NA	$\frac{0}{1260}$	NA	NA		NA	$\frac{0}{1260}$
O9	NA	NA	$\frac{0}{1260}$	NA	$\frac{0}{1260}$	NA	NA	NA		$\frac{0}{1260}$
O10	$+\frac{1575}{1260}$	$\frac{0}{1260}$	$+\frac{2520}{1260}$	$\frac{0}{1260}$	$+\frac{2520}{1260}$	$-\frac{315}{1260}$	$\frac{0}{1260}$	$\frac{0}{1260}$	$\frac{0}{1260}$	

To P = 0.742, P = 0.469 and P = 0.207 levels given in Table 5.13 visual coloring of Table 5.14 corresponds, in which different shading show measures of disagreement of pairwise estimations without considering correcting displacements – f_{ij} value, mentioned in hypotheses that objects importance weights are the weights found at $P = P_f$.

As is obvious in Table 5.13 – beginning with P = 0.1 approximate solutions excursions differ from solution according to spanning tree less than for 1%. This circumstance allowed us to display visually f_{ij} sizes, reduced them to a common denominator.

Excluded at the first stage of algorithm work four pairwise comparisons (from $P = 1.15$ to $P = 0.742$) are painted in dark gray color (Table 5.14). Four pairwise comparisons removed at the second stage (from $P = 0.742$ to $P = 0.469$) are painted in light gray color. Two pairwise comparisons removed at the third stage (from $P = 0.469$ to $P = 0.207$), are marked with shading. The cells of Table 5.14 containing remained pairwise comparisons are marked with double shading. The cells containing pairwise comparisons, compatible with spanning tree edges, are painted in black color.

Let us remark that estimations in table cells with double shading coincide with estimations recalculated through pairwise comparisons' estimating, compatible with spanning tree edges. Actually existence of such "duplicating" communications in pairwise comparisons' matrix as we understand, allows in the course of solution creation through mismatches conflicts elimination to support main preferences framework. Estimations in cells with double shading are rejected at the last stage at P going to zero – for spanning creation.

5.3 Conclusion

The suggested methodology in part 1 and adaptive algorithm for ranking the similar objects are complex by their systemic significance, taken into account the case of ambiguity of the given data on the resource and basic criteria or their partial absence and substantiate a possibility of using the expert approach to the facilities' ranking [17].

To merge the ranked lists of similar objects into a unified ranked list, use has been made of several novelties ensuring correctness of the procedure of comparing the objects of various types on the basis of partial expert findings of equisignificance of their separately selected representative pairs for solution of more general tasks. The suggested approach can be used, for example, by the units in charge of safety of the fuel or energy complex facilities. A complex analysis of interrelated risks for separate industries and the fuel and energy complex as a whole will allow for substantiated recommendations as to determine the required and sufficient safety levels of hazardous production facilities proceeding from their significance for solution of a wide spectrum of management problems [17].

The method, described on part 2, allows to receive compared objects' estimations at incomplete comparisons and giving the chance to follow the structure of fundamental comparisons forming objects importance weights [17].

Thus, suggested method, without denying AHP (method Saaty), allows to expand its opportunities significantly. At program realization of the algorithm it is possible to receive express estimations of compared objects' importance without completing all pairwise comparisons' matrix. The method allows "to control" AHP matrix completing process as it reveals places of critical experts' estimates mismatch which can be more extensive than mismatch of estimates in threes (as usual among many). In concept the method allows to work and with uncoordinated matrixes from which they refuse in AHP, in this case the number of iterations of edges exception from graph will naturally increase significantly [18].

Solution to mismatches conflicts is very similar to Delphi1 method: "ponderability estimation dropping out of general series" – that is on the border – we "cancel" and we try to adapt other estimations to each other in "harder" circumstances [18].

References

[1] Kostrov, A. V. (1996). Interval ranking of objects by many indicators. *Problemy Bezopacnosti Chrezvychainykh Situatsiy* 1, 46–68.

[2] Makhutov, N. A., and Kryshevich, O. V. (2002). Possibility of "man-machine-environment" systems of machine-building profile based on fuzzy sets 2, 94–103.

[3] Gokhman, O. G. (1991). *Expert Assessment.* Voronezh: Voronezh University Publishers, 152.

[4] Zinevich, S. V., Tarasenko, V. A., Usolov, E. B. (2006). "Criteria of objects ranking by emergency hazard levels," in *Proceedings of the 11th Research and Practical Conference*, (Irkutsk: East-Siberian Institute of RF Ministry of Interior), 220–223.

[5] Buyko, K. and Pantyukhova, Y. (2010) Approaches to assessment of industrial safety in organizations operating hazardous industrial facilities. *Occupational safety in industry.* 10, 42–46.

[6] Bruck,V. M., and Nikolayev, V. I. (1977). *Fundamentals of General Systems Theory.* Leningrad: SZPI Publishers.

[7] von Neumann, J., and Morgenstern, O. (2007). *Theory of Games and Economic Behavior (60th Anniversary Commemorative Edition).* Princeton, NJ: Princeton University Press.

[8] Larichev, O. I. (2002). Features of decision making methods in multicriterion individual choice problems. *Avtomatika Telemekhanika* 2, 146–158.
[9] Russman, I. B., Kaplinskiy, A. I., and Umyvakin, V. M. (1991). Modeling and algorithmization of underformalized problems of selecting best system variants. Voronezh: VGU Publishers.
[10] Gokhman, O. G. (1991). *Expert Assessment*. Voronezh: VGU Publishers.
[11] Litvak, B. G. (2004). *Expert Technologies in Management*. Moscow: DELO Publishers, 400.
[12] Cox, D. R., and Hinkley, D. V. (1979). *Theoretical Statistics*, 1st Edn. Boca Raton, FL: Chapman and Hall/CRC, 528.
[13] Kuvshinov, B. M., Shaposhnik, I. I., Shiryayev, V. I., and Shiryayev, O. V. (2002). Use of committees in pattern-classification problems with inexact expert estimations. *J. Comput. Syst. Sci. Int.* 5, 81–88.
[14] Zhukovsky, V. I., and Zhukovskaya, L. V. (2004). *Risk in Multicriterion and Conflicting Systems in Uncertainty*. Moscow: Yeditorial URSS Publishers, 267.
[15] Melikhov, A. N., Berstein, L. S., and Korovin, S. Y. (1990). Situational advising systems with fuzzy logic. Moscow: Nauka Publishers.
[16] Saaty, T. L. (2009). *Mathematical Principles of Decision Making. Principia Mathematica Decernendi*. Pittsburgh, PA: RWS Publications, 531.
[17] Bochkov, A., Lesnykh, V., Zhigirev, N., and Lavrukhin, Y. (2015). Some methodical aspects of critical infrastructure protection. *Saf. Sci.* 29, 229–242. doi: 10.1016/j.ssci.2015.06.008
[18] Bochkov, A. V., and Zhigirev, N. N. (2014). The analytic hierarchy process modification for decision making UNDER UNCERTAINTY. *Reliability* 9, 36–45.
[19] Bradley, R. A. and Terry, M. E. (1952). Rank analysis of incomplete block designs, I. the method of paired comparisons. *Biometrika* 39, 324–345.
[20] Markin, E. G. (1974). *Problem of Multiple Select*. Moscow: Nauka Science, 256.
[21] Kendal, M. (1975). *Grade Correlation*. Moscow: Statistics, 216.
[22] David, G. (1978). *Pairwise Comparison Method*. Moscow: Statistics, 144.
[23] Litvak, V. G. (1982). *Expert Information. Method for Obtaining and Analysis*. Moscow: Radio i Svyaz (Radio and communication), 184.

[24] Saaty, T. L. (2008). *Solution Making at Correspondences and Backlinks: Analytic Nets*. Moscow: Izdatelstvo Press LKI, 360.
[25] Totsenko, V. G. (2002). *Methods and Systems for Supporting of Solution Making. Algorithmic Aspect*. Riev: Naukova dumka, 381.
[26] Harker, P. T. (1987). Incomplete pairwise comparisons in the analytic hierarchy process. *Math. Model.* 9, 837–848.
[27] Cayley, A. (1889). A theorem on trees. *Q. J. Math.* 23, 376–378.

6

Understanding Time Delay Based Modeling and Diffusion of Technological Products

Mohini Agarwal[1], Adarsh Anand[2], Deepti Aggrawal[2] and Rubina Mittal[3]

[1]School of Business, Galgotias University, India
[2]Department of Operational Research, University of Delhi, India
[3]Keshav Mahavidyalaya, University of Delhi, India

Abstract

The diffusion theory and its extended forms have always been an appealing topic in the field of marketing for looking at spread and adoption of a product. Multi-stage nature of diffusion theory has gained little importance in terms of sales estimation via mathematical modeling. Here, in this chapter, we intend to model two-stage diffusion process viz. awareness and adoption by making use of infinite server queuing theory. It is assumed that consumer makes rational use of information achieved by promotional activities in order to reach a decision of purchase. The goal of the present chapter is to investigate the influence of different nature of information flow on the diffusion of a product. Four different models under varied nature of awareness and adoption are modeled and empirically tested on two different consumer durables. The obtained results show good consistency on both datasets.

6.1 Introduction

Queuing theory is the numerical study of waiting lines or queues under different systems. Basically, it attempts to answers: "How long a customer must wait?", "What should be the capacity of a queue?", etc. All of us have experienced the exasperation of having to stand in line for varied reasons [1]. Since, we as customers do not like to wait and also the management of firms does not like us to wait. But reduction of the waiting time usually requires additional investments. To decide whether or not to invest, it is important to know the effect of such investment on waiting time. Thus we need mathematical models and techniques for analysis of such situations. Queuing models are very constructive for these practical problems. Originally this theory was proposed to develop models for predicting the behavior of systems that provide service for randomly occurring demands. These models are successfully employed by researchers in different disciplines in management and reliability estimation in the field of software engineering [2]. Inoue and Yamada (2003) applied infinite server queuing theory to the basic assumption of delayed S-shaped model, i.e. described fault detection process as a two-stage process [3]. Due to its wide-ranging applicability, in the present analysis we attempt to show how to use queuing approach to explain the awareness and adoption process of a product. The underlying idea is awareness process can be looked as an arrival in the queue and adoption process as providing service.

Generally, innovation acts as a survival unit for any organization. Innovation can be defined as something new, better than what already exists in the market, economically viable and should have widespread appeal in the market place. The diffusion process refers to the method of describing about the spread of an innovation in the market [4]. Innovation diffusion process is a decision-making process in which an individual passes from first knowledge to forming an attitude, to a decision to adopt or reject, to implement and finally confirmation [5]. Thus making individuals inform about the features of innovation is crucial for happening of sales. Awareness is defined as a process of receiving and processing information in order to attain perception towards purchase of the product where as adoption describe about individuals decision to utilize an innovation to its fullest [6]. Keeping above discussion in mind we intend to model adoption process being preceded by the awareness flow in the market using unified approach based on infinite queuing theory.

The work in the field of innovation diffusion modeling is immense for describing real life scenarios that may prevail. The earliest external model was proposed by Fourt and Woodlock (1960) [7] which gained popularity after the work of Coleman et al. (1966) [8] that explained the exponential nature of external influence model. After one year, Mansfield (1961)[9] proposed an S-shaped model for explaining the internal influence model. Later, Bass in 1969 proposed mixed influence model, continuing on the same concept of modeling sales [10]; till date there are ample of research paper based on diffusion theory. The research in this field still continues to capture the variability in adoption behavior and other marketing conditions. Multi-stage nature of diffusion theory still remains to be modeled. Thus, considering diffusion process as two-stage process consisting of awareness and adoption process in which some amount of time is consumed in moving from one stage to another. Van den Bulte and Lilien (1997) [11] presented an approach of modeling awareness at separate level than adoption, when one is still bothered about the final sales of the product. Singh et al. (2014) [12] proposed time delay effect innovation diffusion model based on the assumption of time delay between the adoption of product by an innovator and then an imitator. Anand et al. (2014) [13] presented a two-stage modeling framework in which they categorized Product Awareness (*PAWP*) and Product Adoption (*PADP*) as two separate scenarios before the product is eventually bought by the consumers. Their unification methodology was based on an organized study of *PAWP* and *PADP* where product adoption processes are described using awareness process with time delay. Furthermore in 2016, Anand et al. [16] have proposed an unified approach for modeling Innovation Adoption by making use of three stage queuing process for product diffusion.

Applicability of infinite queuing theory in the field of software engineering for determining the fault correction process being dependent on fault detection process led us to make use of this approach in determining the sales when awareness plays a crucial role in happening of sales. In this chapter, a mathematical approach for determining the sales of product when information flow form the basis of customers making purchase of the product is modeled making use of traditional unified infinite queuing theory. Further different probability distribution functions are considered to capture the different trends that are prevailing in the market. The chapter is divided into seven sections where notations are supplemented in Section 2; Section 3 describes in detail the proposed modeling framework; Section 4 consists of data analysis for two different datasets; in Section 5 managerial implication is presented while conclusion and references in Section 6 and 7, respectively.

6.2 Research Methodology

6.2.1 Notations

These are the following set of notations that are used in the modeling:

$N_{Ad}(t)$: Cumulative number of Adopters

$f_{Ad}(t)$; $f_{Aw}(t)$: Probability Density Function for Adoption Process & Awareness Process

$F_{Ad}(t)$; $F_{Aw}(t)$: Distribution Function for Adoption Process & Awareness Process

m : Potential Market Size

The developed methodology is based on infinite server queuing theory (Kapur et al., 2011) which describes about the fact that gap between buyers being informed and adopting the product should not be ignored. Considering, the case when a number of promotional activities for a product are performed in accordance with Non-Homogeneous Poisson Process (NHPP). This eventually results into information flow about the characteristic features of a product. The amount of awareness created at the end of promotional activities form an arrival process. Here, number of sales observed at the end of promotional period is equivalent to number of customers in the $M^*/G/\infty$ queuing system. The arrival process is represented by M^* is an NHPP process and service time has general distribution [2]. Using this theory we characterize adoption process through the awareness process. The proposed modeling framework is based on the following assumptions: Let the counting process $\{X(t), t \geq 0\}$,$\{Y(t), t \geq 0\}$ represents the cumulative number informed individual, number of adopters, respectively, up to time t and the process started at time $t = 0$. Then the distribution of $Y(t)$ is given by:

$$\Pr\{Y(t) = n\} = \sum_{j=0}^{\infty} \Pr\{Y(t) = n \,|X(t) = j\} \frac{\left(N_A(t) e^{-N_A(t)}\right)}{j!} \tag{6.1}$$

If the count for informed individuals is j then the probability that n individuals have adopted via the adoption process is given as

$$\Pr\{Y(t) = n \,|X(t) = j\} = \binom{j}{n} (p(t))^n (1 - p(t))^{j-n} \tag{6.2}$$

where $p(t)$ is the probability that an arbitrary adoption happens at time t, which can be defined using Stieltjes convolution and the concept of the conditional distribution of arrival times, given as:

$$p(t) = \int_0^t F_{Ad}(t-u) \frac{dN_A(u)}{N_A(t)} \tag{6.3}$$

The distribution function of cumulative number of adopters up to time t

$$\Pr\{Y(t) = n\} = \left(\int_0^t F_{Ad}(t-u)\, dN_A(u)\right) \frac{e^{-\left(\int_0^t F_{Ad}(t-u) dN_A(u)\right)}}{j!} \tag{6.4}$$

Equation (6.4) describes that $N(t)$ follows an NHPP with Mean value function (MVF) $(\int_0^t F_{Ad}(t-u)\,dN_A(u))$, i.e.

$$N_{Ad}(t) = \left(\int_0^t F_{Ad}(t-u)\,dN_A(u)\right) \tag{6.5}$$

Hence knowing the MVF for awareness times $N_A(t)$ and distribution of adoption times $F_{Ad}(\bullet)$ we can compute the MVF of a two-stage awareness and adoption process. In order to have deep insight, we have considered different cases to describe the pattern that can be followed by awareness and adoption process; for modeling such scenario we have considered different cases which are explained as follows:

M-1: In this particular case we consider that awareness intensity is constant over time and probability density function for adoption process to be logistic in nature i.e. $N_A(t) \sim 1(t)$ and $f_{Ad}(t) \sim logistic(b,\beta)$. Using Equation (6.7) we get:

$$N_{Ad}(t) = m\left[\frac{1-exp(-bt)}{1+\beta\; exp(-bt)}\right] \tag{6.6}$$

M-2: Assuming that awareness intensity and adoption process follow exponential distribution over time with different rates i.e. $N_A(t) \sim exp(b_1)$ and $f_{Ad}(t) \sim exp(b_2)$. Using Equation (6.7) we get:

$$N(t) = m\left[1 - \frac{1}{(b_1-b_2)}\left(b_1 e^{-b_2 t} - b_2 e^{-b_1 t}\right)\right] \tag{6.7}$$

M-3: Consider awareness intensity to be exponentially changing over time and probability density function for adoption process to be logistic distribution i.e. $N_A(t) \sim exp(b)$ and $F_{Ad}(t) \sim \text{logistic}(b,\beta)$; and again making use of Equation (6.7) we get

$$N(t) = m(1-e^{-bt}) + m(1+\beta)e^{-bt}log\left[\frac{(1+\beta)\,e^{-bt}}{1+\beta e^{-bt}}\right] \tag{6.8}$$

M-4: Assuming that awareness intensity and adoption process as Logistic distribution i.e. $N_A(t) \sim \text{logistic}(b,\beta)$ and $F_{ad}(t) \sim \text{logistic}(b,\beta)$; and again making use of Equation (6.7) we get,

$$N(t) = m\left[\frac{1-e^{-bt}}{1-\beta^2 e^{-bt}}\right] + m\frac{(1+\beta)^2 e^{-bt}}{(1-\beta^2 e^{-bt})^2}\left[bt + 2\log\left[\frac{(1+\beta)\,e^{-bt}}{1+\beta e^{-bt}}\right]\right] \tag{6.9}$$

The above presented five models depict different trend of awareness and adoption that may conquest in market, on which the final sales of the product depends.

6.3 Research Results and Findings

The proposed models have been validated on the sales data of two different cars namely i10 and accent extracted from two online sites (Motorbeam, 2015 [14] and Team-bhp, 2015 [15]). The parameters are estimated using SPSS tool based on non-linear least square method. For the consistency of the model two comparison criteria M.S.E and R^2 are evaluated; where lower value of M.S.E indicate better fit to the data and higher value of R^2 shows better consistency [2].

Tables 6.1 and 6.2 contain the estimated parameter and comparison criterion values for proposed models of diverse conditions that may manifest into growth of any product. Considering M.S.E and R^2 as two main principles for analysis, which implies that M1 performs better in case of both data sets with high value of coefficient of determination and least value of mean squared error. Also, for other models the results are acceptable because of high values of R^2 which is almost tending to one implying better fit to the data.

Figures 6.1 and 6.2 show the graphical representation of proposed modeling framework in contrast to actual of sales of two different products. The graphical view portray the relative scenario among different models which clearly describe that M-1, in which the information flow is constant

Table 6.1 Parameter estimates and comparison criteria of DS-I

#	m	$b\,(b_1)$	b_2	β	R^2	$M.S.E$
M1	390.670	0.0615	—	1.027	0.997	31.146
M2	548.168	0.990	0.021	—	0.994	64.087
M3	349.850	0.128	—	1.023	0.964	373.669
M4	350.230	0.150	—	1.021	0.949	540.544

Table 6.2 Parameter estimates and comparison criteria of DS-II

#	m	$b\,(b_1)$	b_2	β	R^2	$M.S.E$
M1	157.932	0.149	—	1.021	0.997	5.770
M2	180.695	0.99	0.068	—	0.985	28.424
M3	156.870	0.272	—	1.230	0.950	96.604
M4	156.890	0.312	—	1.042	0.932	131.338

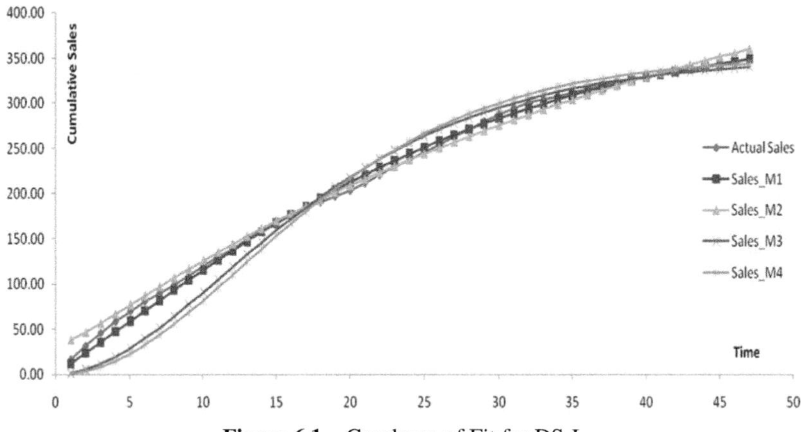

Figure 6.1 Goodness of Fit for DS-I.

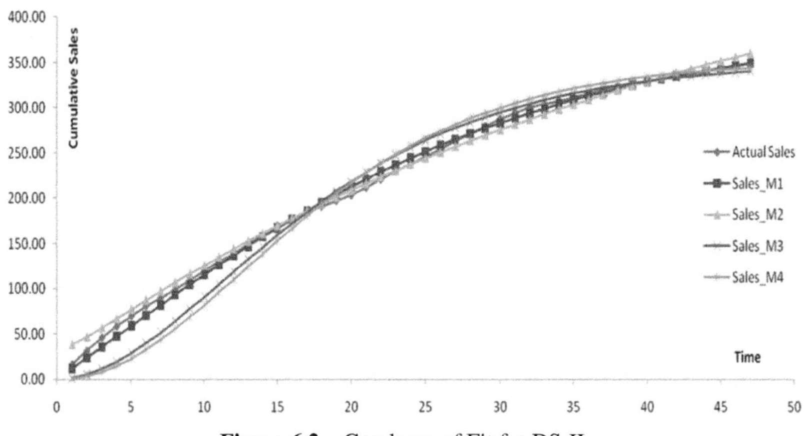

Figure 6.2 Goodness of Fit for DS-II.

and adoption being logistically increasing performs better in distinction with others. This result may vary according to the nature of data being analyzed.

6.4 Discussion

Shift in marketing strategy from being firm oriented to being both firm and customer oriented involves high cost. Thus, there comes the requirement of developing mathematical models for planning and developing products with reduced cost and better risk management for satisfying consumer needs.

The work presented in this chapter attempts to develop a generalized model that takes into consideration two different stages of diffusion process viz. awareness and adoption. Using the infinite queuing theory approach $M^*/G/\infty$ queuing system, we are able to propose a two-stage modeling framework of determining the sales when the manner of information flow has considerable position. The mode of information flow will form the base for developing consumers' perception towards final adoption. This study thus facilitates firms in determining the sales making use of information supplied and thus firms can focus majorly on promotional activities to create a large population of informed individuals who can thus process knowledge received and form their perception to adopt or reject a product.

6.5 Conclusion

The proposed models in this chapter present a new outlook towards estimation of sales based on awareness as an important attribute. Producers are often confused about how to estimate sales when characteristics of product are promoted in different manner among customers. The results estimated in this chapter clearly indicate that sale can be determined based on two-stage modeling framework in a more judicious manner leading to better management of resources. In case of both data sets used model in which awareness is constant and adoption is logistic in nature performs best; also other proposed models are better in their predictability. In future we wish to extend the concept into three-stage modeling framework.

References

[1] Gross, D., and Harris, C. M. (2004). *Fundamentals of Queuing Theory*, 3rd Edn. Hoboken, NJ: Wiley.

[2] Kapur, P. K., Pham, H., Gupta, A., and Jha, P. C. (2011). *Software Reliability Assessment with OR Applications*. London: Springer.

[3] Inoue, S., and Yamada, S. (2003). "A software reliability growth modeling based on infinite server queuing theory," in *Proceedings of the 9th ISSAT International Conference on Reliability and Quality in Design*, 305–309.

[4] Rogers, E. M. (1983). *Diffusion of Innovations*. New York, NY: Free Press.

[5] Anand, A. (2013). *A Study of Innovation Adoption & Warranty Analysis in Marketing and Successive Software Releases*. Ph. D. thesis, University of Delhi, Delhi, India.

[6] Anand, A., Kapur, P. K., Agarwal, M., and Aggrawal, D. (2014a). "Generalized innovation diffusion modeling & weighted criteria based ranking," in *Proceedings of 3rd International Conference on Reliability, Infocom Technologies and Optimization*, Noida, 2014, 1–6. (IEEE Conference Publications). doi: 10.1109/ICRITO.2014.7014705

[7] Fourt, L. A., and Woodlock, J. W. (1960). Early prediction of market success for new grocery products. *J. Mark.* 25, 31–38.

[8] Coleman, J. S., Katz, E., and Menzel, H. (1966). *Medical Innovation: A Diffusion Study*, 2nd Edn. Indianapolis, IN: Bobbs-Merrill.

[9] Mansfield, E. (1961). Technical change and the rate of imitation. *Econometrica* 29, 741–766.

[10] Bass, F. M. (1969). A new product growth for model consumer durables. *Manag. Sci.* 15, 215–227.

[11] Van den Bulte C., and Gary, L. (1997), A Two-Stage Model of Innovation Adoption with Partial Observability: Model Development and Application. *Mark. Sci.* 16, 338–353.

[12] Singh, O., Kapur, P. K., Sachdeva N., and Bibhu, V. (2014). "Innovation diffusion models incorporating time lag between innovators and imitators adoption," in *Proceedings of 3rd International Conference on Reliability, Infocom Technologies and Optimization*, Noida, 2014, 1–6.

[13] Anand, A., Singh, O., Agarwal, M., and Aggarwal, R. (2014b). "Modeling adoption process based on awareness and motivation of consumers," in *Proceedings of 3rd International Conference on Reliability, Infocom Technologies and Optimization*, Noida, 2014, 1–6. (IEEE Conference Publications). doi: 10.1109/ICRITO.2014.7014701.

[14] Motorbeam (2015). Available at: www.motorbeam.com [accessed June 15, 2015].

[15] Team-bhp, www.team-bhp.com, accessed date 15 June, 2015.

[16] Anand, A., Agarwal, M., Aggrawal, D., and Singh, O. (2016). Unified approach for modeling innovation adoption and optimal model selection for the diffusion process. *J. Adv. Manage. Res.* 13, 154–178.

7

Role of Soft Computing in Science and Engineering

Preeti Malik[1], Lata Nautiyal[1] and Mangey Ram[2]

[1]Department of Computer Science and Engineering, Graphic Era (Deemed to be University), India
[2]Department of Mathematics, Computer Science and Engineering, Graphic Era (Deemed to be University), India

Abstract

Soft computing is a blended mixture of various techniques that were intended for modeling the solutions of realistic problems. It is complex to model these problems mathematically. Soft computing can also be defined as an effort to imitate natural creatures like animals, humans which are stretchy, soft, hard, adaptive, rigid and crisp. Hence, soft computing is a set of problem-solving methods that are analogous to biological reasoning and cognitive computing. Fuzzy logic, genetic algorithms, ant colony optimization, neural network and evolutionary algorithms are some of the soft computing techniques. This chapter discusses various soft computing techniques used in science and engineering. This chapter also presents a survey of a range of applications of these fields ranging from the purely theoretical to the most practical ones.

7.1 Introduction

Soft computing is basically an association of methods that works synergistically and provides adaptable information processing means for real life problems [1]. Soft computing is a blended mixture of various techniques that were intended for modeling the solutions of realistic problems. It is complex to model these problems mathematically. Main aim of soft computing is to use the acceptance for ambiguity, uncertainty, and imprecision so as to attain tractability, heftiness and cost-effective solutions [2]. The standard is to develop methods for performing computations that provide a reasonable solution to a formulated problem and the solution should be cost-effective [3]. For example, the ambiguity in a calculated quantity is because of intrinsic variations in the calculation process itself. The ambiguous result is resulted form of the joined and accumulated consequences of these calculated ambiguities used for calculating the result [4].

Soft computing is different from hard (conventional) computing in the way it tolerates ambiguity, uncertainty, imprecision and partial truth. It is based on the notion of human mind. It is basically to optimally solve complex problems. This term was first proposed by Lotfi A. Zadeh as [1]:

> "*Soft computing is a collection of methodologies that aim to exploit the tolerance for imprecision and uncertainty to achieve tractability, robustness, and low solution cost. Its principal constituents are fuzzy logic, neuro-computing, and probabilistic reasoning. Soft computing is likely to play an increasingly important role in many application areas, including software engineering. The role model for soft computing is the human mind.*"

In some problems a lot of effort and cost is needed to improve the level of precision and certainty. Zadeh presented that in travelling sales person problem, computation time is proportional to accuracy and increases exponentially [1]. Soft computing can also be defined as an effort to imitate natural creatures like, animals, humans which are stretchy, soft, hard, adaptive, rigid and crisp. Hence, soft computing is a set of problem-solving methods that are analogous to biological reasoning and cognitive computing. Fuzzy logic, genetic algorithms, ant colony optimization, neural network and evolutionary algorithms are some of the soft computing techniques.

Fuzzy logic is linked to approximation and computing with words and probabilistic reasoning is linked to uncertainty and belief networks. All of these techniques have some common characteristics:

1. These methods are non-linear.
2. These methods follow reasoning paths like human instead of conventional methods.
3. These methods make use of self-learning.
4. These methods exploit yet-to-be-proven theorems.
5. When there is noise or errors then these methods are robust.

According to Kosko, following are some of the similarities between fuzzy logic and neural network [5]:

- approximate functions from test data
- do not require a numerical model
- are dynamic in nature
- can be represented as graphs
- process inaccurate information inaccurately
- have the same solution space
- a set of n neurons defines n-dimensional fuzzy sets
- can perform as an associative memory

Soft computing techniques model the real world problem as a non-linear, time-varying problem. Hard computing needs the accurate mathematical model and huge computations to solve the particular problem [6]. Literature reveals that soft computing techniques have been used in different domains of science and engineering [7, 8] like; industrial engineering, mechanical engineering, etc. Fuzzy logic, evolutionary algorithms, machine learning, probabilistic reasoning, and neural networks are the base of soft computing [9]. These techniques are being applied to aircraft, communication network, power systems and computer sciences, etc. Among these techniques, probabilistic reasoning and fuzzy logic are knowledge-driven techniques and neural network and evolutionary computing are basically data-driven techniques. Figure 7.1 depicts some of the techniques. These techniques use rich knowledge description, knowledge gathering and processing for solving the problems. These three techniques can be deployed as different tools or can also be integrated in.

For optimization problems, evolutionary algorithms are widely used. For prediction tasks, neural network and learning systems are used. Fuzzy logics are basically used for modeling and measuring ambiguities that arise. For some problems more than one technique can also integrate. This chapter presents a theoretical understanding of these techniques in detail.

7.1.1 Why Soft Computing Approach?

Simple systems can be modeled and analyzed by using mathematical methods whereas more complex systems like in biology, science, engineering and medical remain obstinate to mathematical methods. Soft computing deals

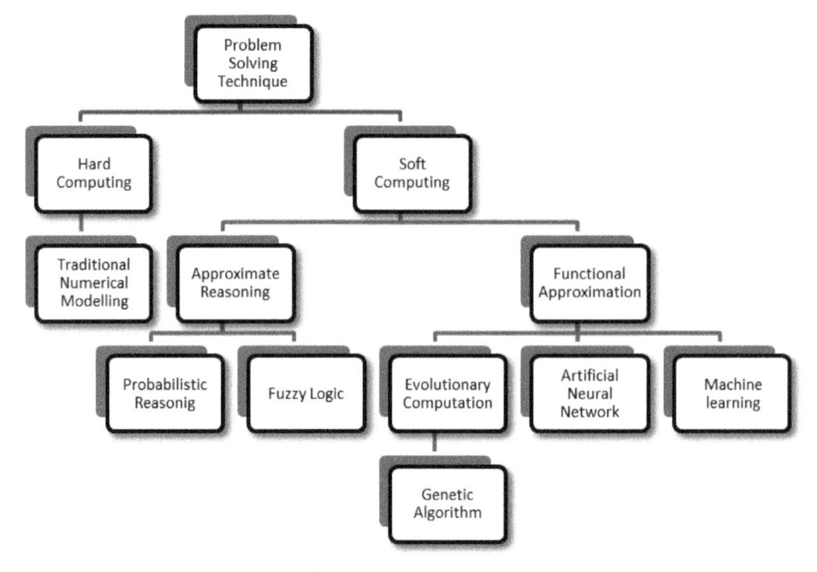

Figure 7.1 Soft computing techniques.

with these complex systems from the discipline of science and engineering. A human being typically has following characteristics:

1. He can take decisions.
2. He can infer from the previous knowledge.
3. He adapts the changes in the environment.
4. He learns for his betterment.

Some of the characteristics of soft computing methods are:

1. Human expertise
2. Biologically motivated computing models
3. Novel optimized methods
4. Novel application domains
5. Model-free learning
6. Fault tolerance
7. Ambiguity handling

7.2 Soft Computing Techniques

Some of the soft computing techniques are:

- Machine learning
- Fuzzy logic (FL)

- Evolutionary algorithms
- Genetic algorithms
- Bayesian network
- Neural network
- Particle swarm optimization
- Ant colony optimization

7.2.1 Machine Learning

Alpaydin [10] defined machine learning as:

> *"Optimizing a performance criterion using example data and past experience"*

An essential role is played by data in machine learning and then learning algorithms are applied to this data to discover useful pattern and properties. Performance of learning algorithms is affected by the quantity and quality of data. Figure 7.2 depicts an example of the two-class dataset.

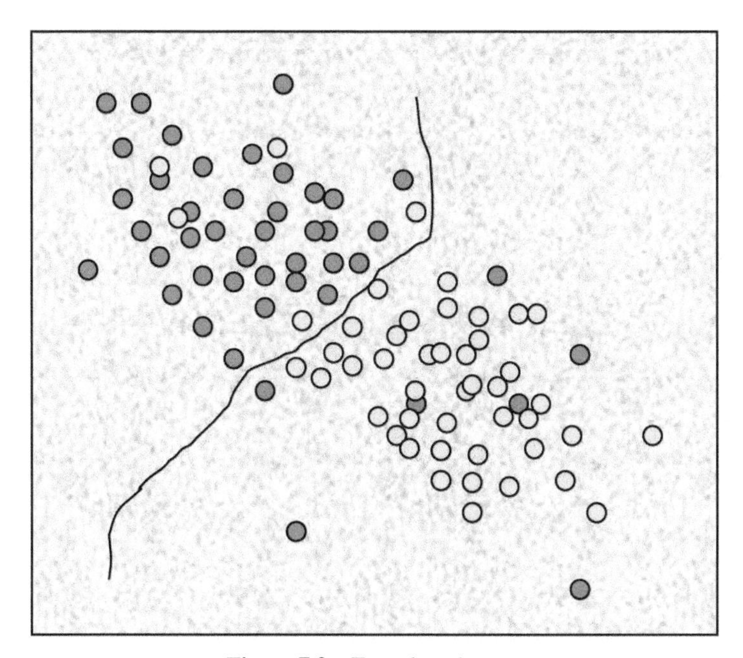

Figure 7.2 Two-class dataset.

7.2.1.1 Notation of dataset

The dataset can be of two types; labeled and unlabeled.

- **Labeled Dataset:** X=$\{x^{(n)} \in \mathrm{R}^d\}_{n=1}{}^N$, Y=$\{y^{(n)} \in \mathrm{R}^d\}_{n=1}{}^N$
- **Unlabeled Dataset:** X=$\{x^{(n)} \in \mathrm{R}^d\}_{n=1}{}^N$

where, X is the **feature set** that contains N samples. Each sample is a d-dimensional vector $x^{(n)}$=[$x_1{}^{(n)}$, $x_2{}^{(n)}$, $x_3{}^{(n)}$......$x_d{}^{(n)}$] and this sample is called feature vector or feature sample. And each $x_i{}^{(n)}$ is called attribute, feature or element. Y is **label set**. Some application doesn't make use of label set.

7.2.1.2 Training data and test data

Machine learning assumed that a universal dataset already exists and this dataset consists of maximum possible pair of data, probability distribution in real world whereas in real life application due to the shortage of memory and other resources, it is observed that only a subset of universal dataset is available. This data is known as training data because it is used to learn patterns and knowledge of the data. In addition to learning from data, machine learning also require some kind of prediction based on the previous knowledge. This prediction is done to check the performance of learning and the dataset used for this purpose is called **test data**. Additionally, the two critical factors of machine learning are **modeling** and **optimization**. Modeling is concerned with separating boundaries of the given training set, and then optimization techniques are used to search the best parameters of the model that is selected. In Figure 7.3, an explanation of the three datasets given above is presented.

7.2.1.3 Relationships with other disciplines

Machine learning is also linked to other fields like data mining, artificial neural networks, information retrieval, pattern recognition, artificial intelligence, function approximation, etc. Machine learning focuses more on learn, model, optimization, and regularization of the data than these disciplines so as to make best use of the data. Machine learning is based on both statistics and computer science:

- **Statistics:** Statistical properties are learnt and inferred from the data.
- **Computer science:** Optimization algorithms, model representation and performance evaluation.

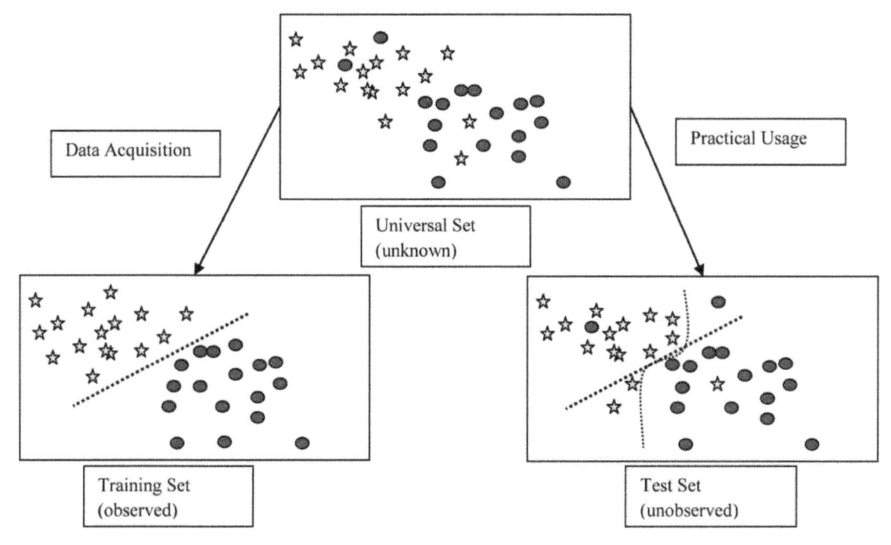

Figure 7.3 Three labeled dataset.

7.2.1.4 Basic concepts and ideals of machine learning

There are situations that occur in daily life when we have to make decisions like if there is cloudy sky outside then we have to bring umbrella. For making machines capable of taking this type of decisions we model the problems in term of mathematical models. We can design the mathematical expression directly from the problem background, for example a vending machine can detect false currency by comparing it with the standards and decoration of the currency. Learning methods are used to relate these two things.

7.2.1.5 The categorization of machine learning algorithms

Machine learning algorithms can be classified into three learning algorithms based on the problem and provided data set:

- **Supervised learning algorithm**: The task of supervised learning infers a function from labeled training data [11]. There is a collection of training examples and each example is a pair that consists of input object and desired output value (supervisory signal). In supervised learning algorithm, training data is analyzed and a function is inferred from the analyses then this inferred function is used mapping new data. In optimized scenario, the algorithm correctly determines the label for unknown data. For this, the algorithm generalizes from training data to unknown data in a 'reasonable' way.

- **Unsupervised learning algorithm**: The task of supervised learning infers a function that describes hidden information or structure from unlabeled data. In unsupervised learning there is no way to measure the accuracy of the results. Unsupervised learning solves problems like summarizing and analyzing the key features of the dataset. Some of the approaches of unsupervised learning are:
    - clustering
    - anomaly detection
    - neural networks
    - approaches for learning latent variable models such as
- **Reinforcement learning algorithm**: This category of machine learning is encouraged by behaviorist psychology. These algorithms are worried about how software agent is supposed to take *actions* in order to maximize view of increasing *reward* in an environment. It is applied in number of applications like game theory, operational research, multi-agent systems, etc. In reinforcement learning accurate input–output pairs are never presented. This type of algorithm is concerned with the on-line performance of the algorithm and a balance between exploration and exploitation [12].

7.2.2 Fuzzy Logic

To deal with the concept of linguistic variables Lotfi A. Zadeh [13], a professor at University of California, proposed a mathematical tool called fuzzy set theory. Fuzzy logic allows in-between values to be defined and it is multi-valued logic basically. The things were started long ago, Parminedes suggested the foremost version of the law (400 B.C.). Plato laid the base and that base became fuzzy logic. He presented that there was also a third are (beyond true and false) where the conflicts "tumbled about".

7.2.3 Evolutionary Algorithms

Evolutionary algorithms are population-based metaheuristic optimization algorithms and a subset of evolutionary computation. It uses the concepts of biological evolution like selection, mutation, recombination and reproduction. Candidate solutions are the individuals of the population and fitness functions are used to measure the quality of the solution. Population is then evolved by repeatedly applying the above operations.

7.2.3.1 Implementation

1. Randomly generate initial population.
2. For each individual, they evaluate the fitness in the population.
3. Repeatedly perform the following computation until termination condition is met:
   3.1 Choose the best-fit individuals for reproduction.
   3.2 Apply crossover and mutation operations to selected individuals.
   3.3 Apply fitness function on new individuals.
   3.4 Drop least-fit individuals and include new individuals.

7.2.3.2 Types

Similar techniques differ in genetic representation and other implementation details, and the nature of the particular applied problem.

- Genetic algorithm
- Genetic programming
- Evolutionary programming
- Gene expression programming
- Evolution strategy
- Differential evolution
- Neuro-evolution
- Learning classifier system

7.2.4 Genetic Algorithms

In genetic algorithms, the population of individuals is progressed towards for the betterment of solution. Each individual has some properties (also called chromosomes or genotype) and these properties can be operated on. Generally, the solution is represented in binary form (in form of 0's and 1's) [14]. Genetic algorithms are iterative in nature and start with randomly generated individuals. The population at each iteration is called *generation*. At each iteration, the fitness function is applied to each individual and more fit individuals are selected for the current population and each individual is mutated. The fitness of an individual is the value of the objective function of the optimization problem. The new generation is used for the next generation. The algorithm stops when either the number of iteration reaches its limit or a required fitness level is reached. A standard genetic algorithm needs:

1. a genetic representation of the solution domain and
2. a fitness function to evaluate the solution domain.

Once the foundation of the algorithms is complete (gene representation and fitness function), the algorithm then initializes the population and applies mutation, crossover and selection operations repeatedly to improve the solution of the problem being solved.

7.2.4.1 Initialization

Size of the population depends on the problem but generally it is hundreds or thousands. Generally, the initial population is randomly generated which allows to cover the entire search space but sometimes the solution may be seeded where there is a good chance of finding the optimal solution.

7.2.4.2 Selection

In each iteration some individuals are selected for breeding a new generation. The individuals are selected based on their fitness value; it means the probability of selecting fitter individuals is more. There are various methods to select the individuals that rate the fitness value of individuals. Some methods rate only random individuals of the entire population because the earlier process is a time-taking process.

The fitness function is defined to measure the quality of the solution and depends on the problem being solved. For example, knapsack problem is to fill the knapsack so as to maximize the profit by placing an object in it. The solution can be represented by an array of bits where 0 means the object is not in the knapsack and 1 represents an object is in the knapsack. Some of these representations are not valid because the size of the knapsack is limited. The sum of values of objects in the knapsack is defined as the fitness of the solution, only if it is valid, i.e. the sum should be less than the limit of the knapsack otherwise 0. It is even hard to define the fitness function in some complex problems; in such situation a simulation is used for fitness function.

7.2.4.3 Genetic operators

Next step is to apply crossover and mutation operations on the selected individuals to generate next generation. For each new individual, a pair is selected to breed and a new individual is produced by applying crossover and mutation operation. This pair is called the *parents*. This process is repeated until a new generation of suitable size is generated. Some researches proposed that using more than two parents for reproduction produce better quality individuals [15, 16].

Belief is separated more importance of crossover against mutation. Fogel [17] supports mutation-based search. Even though these two operators are known as the chief genetic operators, there are other operators such as regrouping, colonization–extinction, or migration [18].

7.2.4.4 Termination

The process of generating new individuals is repeated until a termination condition is reached. Some common stopping conditions are:

- An individual is found that suits minimum criteria.
- A preset number of generations achieved.
- Preset budget reached.
- The solution has reached a plateau, i.e. following iterations no longer make better results.
- Manual check.
- Some combinations of the above cases.

7.2.5 Bayesian Network

Bayesian network is a type of statistical model that depicts random variables and dependencies among those variables. Dependencies are presented by directed cyclic graphs (DAG). Bayesian network is used in healthcare to analyze the relationship between symptoms and disease. It can be used to calculate the probability of occurrence of a disease. In graphs, the nodes represent the random variables and these random variables may be some observable quantities, some unknown parameters, etc. Dependencies are represented by edges; if there is no edge between two nodes it means that those two random variables are conditionally independent. Each node is associated with a probability function that takes, as input, a particular set of values for the node's parent variables, and gives (as output) the probability (or probability distribution, if applicable) of the variable represented by the node. Sequence variables can also be modeled by using the Bayesian network and such type of Bayesian networks are called dynamic Bayesian network. Generalized Bayesian networks are called influence diagrams.

7.2.6 Neural Network

Neural networks inherit the idea from biological neurons. Neurons are the simplest processing units of brains and there thousands or millions of neurons work in a parallel manner. It is accepted as true that intelligence in human results from these neurons [2]. These networks are driven by the information not by the data [19]. The network in divided into layers; input and output

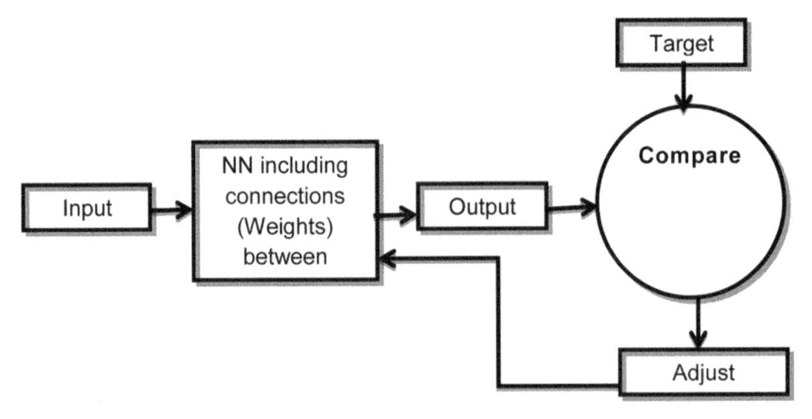

Figure 7.4 Neural network.

and some in-between layer may also exist there. Back propagation network is a type of neural network, where there is an input layer, an output layer and one or more hidden layers [20]. Training is performed by adjusting the values of weights between elements. These networks are trained so as to map a particular input to a suitable output (see Figure 7.4) [21]. There is a number of training algorithms. Back Propagation algorithm adjusts the weights by calculating the difference between the actual output and the desired output [20].

7.2.7 Particle Swarm Optimization

Along with the search process Particle Swarm Optimization (PSO) algorithms also introduce the concept of memory. Unlike genetic algorithms PSO maintains information about the best solution. Both best solutions, local and global are stored in memory. Kennedy and Eberhart [22] first developed the PSO algorithms to simulate the social behavior of bird flocking and fish schooling. It maintains quite a lot of candidate solutions and at each iteration, each individual is evaluated. Each candidate is considered as a particle flying through the fitness setting to find optimal objective function. Each particle stores its information like fitness value, velocity and best fitness value achieved so far. At last, PSO algorithm stores the best fitness values among all particles in the swarm. Following are the steps of PSO algorithm that are repeated until the terminating condition is met:

1. Initialize all the three parameters; population, location and velocity.
2. Assess the fitness of the individual particles (Pbest).

3. Repeat step until terminating criteria met:
   3.1 Store the information of the individual with the highest fitness (Gbest).
   3.2 Adjust velocity according to Pbest and Gbest location.
   3.3 Revise the position of the particle.

The flowchart of PSO algorithm is shown in Figure 7.5.

7.3 Applications

A number of real-world applications are using Soft computing techniques like signal processing, healthcare, disease diagnostic systems, quality assurance system, pattern recognition, robotics, natural language processing, speech

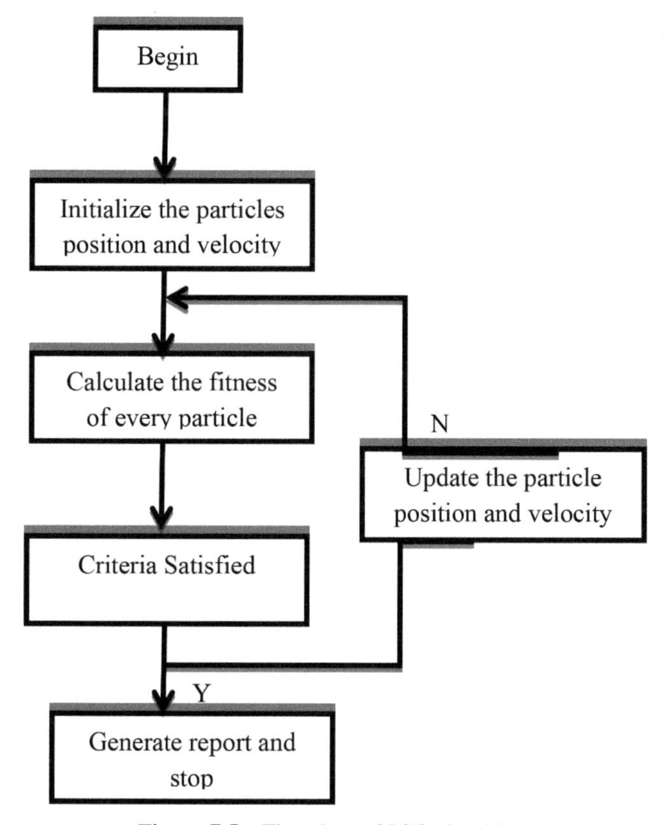

Figure 7.5 Flowchart of PSO algorithm.

processing and much more [1]. For better result, more than one technique can be combined together and this becomes a trend now days to make hybrid systems. Neuro-fuzzy architecture is an example of this kind of systems. In this type of systems fuzzy system is used to derive rule and neural networks induce the rules. To increase the performance fuzzy logic may be employed. Fuzzy logic may control the vibration of direction for a searching vector in quasi Newton method [22]. There is a number of research books and journal articles concentrate on the applications of soft computing [23–36]. The industry is also getting benefited by accepting these techniques. A variety of problems that are hard to solve can easily be solved by these techniques like multi-variable, probabilistic reasoning, etc.

The applications cover the span from the merely hypothetical ones, those which expand new lines in theoretical mathematics, passing across the fields of multimedia, information retrieval, image processing, hybrid intelligent systems, etc. to realistic applications areas like robotics and industry, medical engineering.

Pure and applied mathematics: Hypothetical basics of soft computing techniques branch from merely mathematical ideas. The essential mathematical formalisms of soft computing and fuzzy logic have activated a transformed concern in some old practices, like the theory of *t*-norms and copulas, and have started a total revamp of well found fields like theory of differential equations (fuzziness added), topology, development and arithmetical study of novel logical systems to deal with imprecision, ambiguity and uncertainty [37].

Extended tools for fuzzy learning: There is a number of tools available that are extended for knowledge representation and reasoning like prolog-based implementation. It is extended for lattice-valued logics [38, 39]. Some other methods also comprise the adaptation of improvements and precise optimization methods, like tabulation methods for logic programming.

Multimedia processing: Soft computing techniques are strong at their ability to learn and reason, therefore they can easily be applied to the field of multimedia processing. In this field, the soft computing techniques can be used in image analysis, video sequence, biometric application, image segmentation and color quantization, etc. This is give rise to the use of soft computing in commercial, industrial and military applications [40].

Decision making: Standard methods of decision making give equal preference to all the information but this is not the criteria in real life applications. Therefore, fuzzy logic approaches are applied to the process of decision making. Some particular application areas of preferences modeling are: data-base theory, categorization and data mining, information retrieval, recommendation systems, etc. [41].

Ontologies and the semantic web: Fuzzy logics are being used to fill the gap between intuitive knowledge and machine-readable knowledge. Various methods have been proposed by researchers to extract incomplete and uncertain knowledge and also described the way to handle ambiguity when representing the information. Semantic webs require maintaining a large amount of fuzzy information. Automatic tools are there for reasoning the fuzzy dependencies [42].

Business: Soft computing techniques can also be used in decision making environments like business and economics to deal with the imprecision of human thought and intricacies in estimating inputs. There are surplus applications in business and economics which cover the span from marketing to finance, e-business etc.

Healthcare: Successful diagnosis depends on the knowledge and ability of the inspector but this dependency is risky and not worthy. Giving knowledge to the beginners is a very tedious task because the ability to diagnose the feeling is based on subjective evaluation. Therefore, healthcare industry requires efficient engineering methodologies to assess information. Soft computing is one of the methodologies. Soft computing is used in diagnosing of disease. These technologies relate the symptoms of disease with the disease itself. Sometimes early diagnosis may provide better control on the disease. Soft computing technologies are being used in information radiography, image processing, etc.

Information retrieval: These types of systems are basically concerned with fast and efficient access to a large amount of information. Soft computing methods are being used to model bias and to provide an adaptive environment for information retrieval. This adaptive environment learns from the user's preferences. This modeling can be done by fuzzy logic, probabilistic reasoning or rough sets, etc. Thus soft computing techniques provide greater flexibility to information retrieval systems.

Fuzzy control applications: Fuzzy logics were first applied to the control application for regulating the steam engine and this is proposed by Mamdani and Asilian [43]. After that, the use of fuzzy logic increased rapidly. Soft computing techniques overcome the problem of complexity of control applications and also provide the tolerance ff imprecision, ambiguity, etc.

Robotics: There is a number of sub-fields that make use of soft computing techniques. For example, movement of a robot can be controlled by a neuro-fuzzy system that produces *action* commands for the motor. Information about the surrounding works as the input to this system and this information is gathered from the vision system and goal spotting device. For performing learning, neural networks are used.

7.4 Conclusion

Soft Computing is a blended mixture of various techniques that were intended for modeling the solutions of realistic problems. It is complex to model these problems mathematically. The main aim of soft computing is to use the acceptance of ambiguity, uncertainty, and imprecision so as to attain tractability, heftiness and cost-effective solutions. Machine learning, neural networks, evolutionary algorithms, genetic algorithms, etc. are some major soft computing techniques. More than one technique of soft computing may be merged to provide a better solution to real life problems. This chapter briefly recalls the basics of soft computing techniques. This chapter also presents a survey of a range of applications of these fields ranging from the purely theoretical to the most practical ones.

References

[1] Zadeh, L. A. (1994). "Fuzzy logic, neural networks, and soft computing," in *Proceedings of the Communications of the ACM*, Vol. 37, (New York, NY: ACM), 77–84.

[2] Mitra, S., and Acharya, T. (2003). *Data Mining Multimedia, Soft Computing, and Bioinformatics*. Hoboken, NJ: John Wiley & Sons publication.

[3] Pal, S. K., and Mitra, S. (1999). *Neuro-Fuzzy Pattern Recognition: Methods in Soft Computing.* New York, NY: John Wiley & Sons.

[4] Kirkpatrick, L. D., and Wheeler, G. F. (1992). *Physic. A World View.* New York, NY: Saunders College Pub.

[5] Kosko, B. (1992). "Fuzzy systems as universal approximators," *Proceedings of the First IEEE Conference on Fuzzy Systems*, San Diego, 1153–1162.

[6] Kurhe, A. B., Satonkar, S. S., Khanale, P. B., and Ashok, S. (2011). Soft computing and its applications. *BIOINFO Soft Comput.* 1, 5–7.

[7] Liao, S. H. (2005). Expert system methodologies and applications – a decade review from 1995 to 2004. *Expert Syst. Appl.* 28, 93–103.

[8] Dote, Y., and Ovaska, S. J. (2001). Industrial applications of soft computing: a review. *Proc. IEEE* 89, 1243–1265.

[9] Das, S., Kumar, A., Das, B., and Bumwal, A. P. (2013). On soft computing techniques in various areas. *Comput. Sci. Inf. Technol.* 3, 59–68.

[10] Alpaydin, E. (2010). *Introduction to Machine Learning*, 2nd Edn. Cambridge, MA: The MIT Press.

[11] Mohri, M., Rostamizadeh, A., and Talwalkar, A. (2012). *Foundations of Machine Learning.* Cambridge, MA: The MIT Press. ISBN9780262018258.

[12] Kaelbling, L. P. Littman, M. L., and Moore, A. W. (1996). Reinforcement learning: a survey. *J. Artif. Intell. Res.* **4**, 237–285.

[13] Zadeh, L. A. (1965). Fuzzy sets. *Inform. Control* 8, 338–354.

[14] Whitley, E. (1994). "Spreadsheet module manual for training in research and innovation management," in *Proceedings of the Second Course on Research and Innovation Manageme,* ed. F. Augusto (World Scientific Publishing Co Pte Ltd, London), 66.

[15] Eiben, A. E. et al. (1994). "Genetic algorithms with multi-parent recombination," in *Proceedings of the Third Conference on Parallel Problem Solving from Nature International Conference on Evolutionary Computation,* 78–87.

[16] Ting, C.-K. (2005). On the mean convergence time of multi-parent genetic algorithms without selection. *Adv. Artif. Life* 3630, 403–412. ISBN 978-3-540-28848-0.

[17] Fogel, D. B. (ed.) (1998). *Evolutionary Computation: The Fossil Record.* New York, NY: IEEE Press. *ISBN 0-7803-3481-7.*

[18] Akbari, Z. (2010). A multilevel evolutionary algorithm for optimizing numerical functions. *Int. J. Ind. Eng. Comput.* 2, 419–430.

[19] Jain, L. C., and Martin, N. M. (1998). *Fusion of Neural Networks, Fuzzy Systems and Genetic Algorithms: Industrial Applications*. Boca Raton, FL: CRC Press.

[20] Dhingra, A. K. (2011). *Multi-Objective Flow Shop Scheduling Using Metaheuristics*. Ph.D. thesis, NIT, Krukshetra.

[21] Ballm, R., and Tissot, P. (2006). "Demonstration of artificial neural network in matlab", *Matlab R2006b Help, MathWorks*, Texas A&M University, Corpus Christi.

[22] Kennedy, J., and Eberhart, R. (1995). "Particle swarm optimization," in *Proceedings of the IEEE International Conference on Neural Networks*. (Piscataway, NJ: IEEE Press), 1942–1948.

[23] Kawarada, H., and Suito, H. (1996). *Fuzzy Optimization Method*. Hoboken, NJ: John Wiley & Sons.

[24] Aliev, R. A., Fazlollahi, B., and Aliev, R. R. (2004). *Soft Computing and its Applications in Business and Economics*. Heidelberg: Springer.

[25] Castillo, O., Melin, P., Kacprzyk, J., and Pedrycz, W. (eds) (2008). *Soft Computing for Hybrid Intelligent Systems, Studies in Computational Intelligence,* Vol. 154. Heidelberg: Springer.

[26] Chiclana, F., Herrera-Viedma, E., Alonso, S., and Herrera, F. (2008). "Special issue on fuzzy approaches in preference modelling, decision making and applications," in *Proceedings of the International Journal of Uncertainty, Fuzziness and Knowledge-Based Systems*, (River Edge, NJ: World Scientific Publishing), 16(2 supp).

[27] Dillon, T. S., Shiu, S. C., and Pal, S. K. (2004). Soft computing in case based reasoning. *Appl. Intell.* 21, 231–232.

[28] Ding, L. (2001). *A New Paradigm of Knowledge Engineering by Soft Computing*. Singapoore: World Scientific.

[29] Egiazarian, K., and Hassanien, A. E. (2005). Special issue on soft computing in multimedia processing. *Informatica* 50, 251–252.

[30] Herrera-Viedma, E., and Cordón, O. (2003). Special issue on soft computing applications to intelligent information retrieval on the internet. *Int. J. Approx. Reason*. 34, 89–95.

[31] Hou, Z.-G., Polycarpou, M. M., and He, H. (2008). Special issue on neural networks for pattern recognition and data mining. *Soft Comput.* 12, 2008.

[32] Jamshidi, M., Barak, D., Baugh, S., Vadiee, N., (1993), "Computational and experimental environments for fuzzy logic and control", *Comput. Electr. Eng.*, 19, 289–298

[33] Kercel, S. W. (2006). Industrial applications of soft computing. *IEEE Trans. Syst. Man Cybern.* 36, 450–452.

[34] Langari, R., and Yen, J. (2001). Fuzzy logic at the turn of the millennium. *IEEE Trans. Fuzzy Syst.* 9, 481–482.

[35] Ma, Z. (ed.) (2006). *Soft Computing in Ontologies and Semantic Web.* Heidelberg: Springer.

[36] Nachtegael, M., Kerre, E., Damas, S., and van der Weken, D. V. (2009). Special issue on recent advances in soft computing in image processing. *Int. J. Approx. Reason.* 50, 1–2.

[37] Torra, V., and Narukawa, Y. (2008). "Special issue on soft computing methods in artificial intelligence," in *Proceedings of the International Journal of Uncertainty, Fuzziness and Knowledge-Based Systems.* (River Edge, NJ: World Scientific Publishing), 16(1 supp).

[38] Medina, J. (2009). "Overcoming non-commutativity in multi-adjoint concept lattices," in *Proceedings of the IWANN'09*, Raleigh, NC.

[39] Julián, P., and Rubio-Manzano, C. (2009). "A similarity-based wam for bousi-prolog," in *Proceedings of the IWANN'09*, Raleigh, NC.

[40] Munoz-Hernandez, S., Pablos Ceruelo, V., and Strass, H. (2009). "Rfuzzy: an expressive simple fuzzy compiler," in *Proceedings of the IWANN'09*, Raleigh, NC.

[41] Lopez-Molina, C., Barrenechea, E., Bustince, H., Couto, P., Baets, B. D., and Fernández, J. (2009). "Edge detection based on gravitational forces," in *Proceedings of the of IWANN'09*, Raleigh, NC.

[42] Julián, P., Moreno, G., and Penabad, J. (2009). "On the declarative semantics of multiadjoint logic programs," in *Proceedings of the IWANN'09*, Raleigh, NC.

[43] Cordero, P., Enciso, M., Mora, A., and de Guzmán, I. (2009). "A complete logic for fuzzy functional dependencies over domains with similarity relations," in *Proceedings of the IWANN'09*, Raleigh, NC.

[44] Mamdani, E. H., and Assilian, S. (1975). An experiment in linguistic synthesis with a fuzzy logic controller. *Int. J. Man Mach. Stud.* 7, 1–13.

8

Complex System Reliability Analysis and Optimization

Anuj Kumar[1], Sangeeta Pant[1] and Mangey Ram[2]

[1]Department of Mathematics, University of Petroleum and Energy Studies, India
[2]Department of Mathematics, Computer Science and Engineering, Graphic Era (Deemed to be University), India

Abstract

This chapter provides the various aspects related to reliability analysis and reliability optimization by presenting a brief introduction of reliability, reliability measuring parameters, stochastic processes, Copula method and reliability optimization followed by a review of the literature. The supplementary variable technique, birth–death process, multi-objective particle swarm optimization, mathematical model and block diagram associated with reliability modeling have also been discussed. Further, the reliability–cost optimization of complex bridge system is discussed.

8.1 Introduction

Almost every one of us is using the term reliability in their routine life. When we say the system is reliable it means that the system will perform its task adequately for a reasonable time period.

Reliability is the ability of a system to work properly within the established limits under given operating conditions. In other way reliability can be defined as a function of environment as well as the component itself and in particular depends upon environmental variability.

Reliability is the probability of performing specified function without failure under given conditions for a specified period. Quantitatively, reliability can be viewed as the probability of a system that it will work properly as desired. Quantitatively, reliability is given by

$$R(t) = P(T > t) \qquad 0 \leq t < \infty \tag{8.1}$$

where T denotes the time of failure of the device and t is any time.

Generally, one can assume that the device is working normally initially and every device definitely suffers failures. Therefore,

$$R(0) = 1 \quad \text{and} \quad \lim_{t \to \infty} R(t) = 0. \tag{8.2}$$

Furthermore, $R(t)$ is non-increasing (at most constant, generally decreasing) function for $[0 \leq t < \infty]$.

8.1.1 Reliability Measuring Parameters

(a) **Point-wise availability:** It can be termed as the probability that the system will be able to operate within tolerance limits at a given instant and is also called operational readiness. Symbolically point-wise availability is $A(t) = P[X(t) = 1]$ where $X(t)$ is a binary variable having value 1 and 0, respectively, for the operation and non-operation of the system at an instant t.

(b) **Interval availability:** It is defined as the expected fraction of a given interval of time that the system will be able to operate within tolerances. It is also called the efficiency of the system and its limiting value is the inherent availability.

(c) **Steady-state or asymptotic availability:** It is defined as the probability that in the long run, system operates satisfactorily. Symbolically, the steady-state availability is

$$A(\infty) = \lim_{t \to \infty} A(t)$$

(d) **Interval reliability:** It is the probability of a system that at a specified time t it is operating and remains operative for the duration t to $t + x$. symbolically, interval reliability is

$$R(t, x) = P[X(u) = 1, t \leq u \leq t + x] \tag{8.3}$$

(e) **Instantaneous failure rate [$r(t)$] (also called hazard rate):** It is defined as the limit of the failure rate when the interval length approaches zero.

(f) **Meantime to failure (MTTF):** The expectation of a random variable T is called mean time to system failure (MTSF) or some time to first failure (MTFF) or simply mean time to failure. Mathematically it is defined is

$$MTTF \quad \text{or} \quad MTFF = E(t) = \int_0^\infty R(u)du \tag{8.4}$$

Where *R*(*t*) is the reliability function of the random variable *T* at time *t*. If *R*(*s*) is the Laplace transform of *R*(*t*). Then

$$MTTF = \lim_{x \to 0} \bar{R}(s) \tag{8.5}$$

(g) Mean sojourn time: Mean sojourn time or mean survival time of a system is the expected time which has been taken by it in a particular state before transiting to any other state. If *t* be the sojourn time in any state S_i say, then the mean sojourn time in state S_i is

$$\mu_i = \int_0^\infty P(T > t)dt \tag{8.6}$$

(h) Expected number of repairs by repair facility: Let *V*(*t*) be the random variable representing the number of repairs by repair facility in interval (0, *t*], then the expected number of repairs per unit time is given by

$$E(v) = \lim_{t \to 0} \frac{E\{V(t)\}}{t} \tag{8.7}$$

8.1.2 Stochastic Processes

When the failure and repair time distribution are governed by exponential probability, it becomes essential to study some characteristics of stochastic process. The process, which changes with time in manner, which involves probability, is known as stochastic process. This type of process is further divided into two types, i.e. be Markovian process and non-Marcovian process. A process is said to be Markovian, if the present knowledge is adequate to predict its future behavior otherwise it termed as non-Marcovian.

8.1.3 Copula Method

The copula method has been developed by Sklar in 1959. He showed that any joint distribution function *F* can be seen as a copula function. Different families of copulas have been proposed and these have been described by Nelson (1999) [1]. For most multivariate environmental data the dependence structure may be represented by an Archimedean copula.

An Archimedean copula can be written in the following way; the joint multivariate distribution function

$$C(u_1, u_2, \ldots\ldots\ldots, u_m) = \phi^{-1}[\phi(u_1) + \phi(u_2) + \ldots\ldots\ldots\ldots\ldots + \phi(u_m)] \tag{8.8}$$

For all $0 = u_1, u_2, ..., u_m = 1$. Where ϕ is a function termed as generator; satisfying ϕ (1) = 0.

In order to express one parameter Archimedean copula for two random variables X and Y with there CDFs, respectively as $F_X(x)$ and $F_Y(y)$, let $U = F_X(x)$ and $V = F_Y(y)$. Then U and V are uniformly distributed random Variable. One parameter Archimedean copula denoted as C_θ, can be expressed as:

$$C_\theta(u, v) = \phi^{-1}\{\phi(u) + \phi(v)\}, 0 < u, v < 1 \tag{8.9}$$

where u is the specific value of U and v is the specific value of V, $\phi(.)$ is the copula generating function. This Archimedean copula representation permits to express a multivariate formulation in terms of a single univariate function. In this study, Gummble-Hoggard family is employed [2].

The Gummble-Hoggard family is expressed as:

$$\begin{aligned} C_\theta(u, v) &= C_\theta[F_X(x), F_Y(y)] = H_{X,Y}(x, y) \\ &= \exp[-\{(-\ln u)^\theta + (-\ln v)^\theta\}^{1/\theta}] \end{aligned} \tag{8.10}$$

where, $\theta \in [1, \infty).\theta$ is the parameter of the generating function. And

$$\phi(t) = (-\ln t)^\theta \tag{8.11}$$

with $t = u$ or v, a value of a uniformly distributed random variable varying from 0 to 1. It should be noted that $(-\ln u)^\theta = \phi(u)$ and $(-\ln v)^\theta = \phi(v)$. Parameter θ is related to Kendall's coefficient of correlation τ, between X and Y as:

$$\tau = 1 - \theta^{-1} \tag{8.12}$$

8.1.4 Reliability Optimization

Optimization is a field of applied mathematics that deals with finding the extreme (optimal) value of a function in a domain of definition, subject to various constraints on the variable values. The applicability of optimization methods is widespread, reaching into almost every activity in which numerical information is processed (science, engineering, mathematics, economics, commerce, etc.). Optimization of system reliability plays a vital role to develop optimal system design architecture. System reliability optimization problems have been solved by using many optimization techniques including meta-heuristics [3–5]. In this chapter, the applicability of multi-objective evolutionary algorithm (MOEA), namely the Multi-objective particle swarm optimization in solving the reliability optimization problem of a complex system with two conflicting objectives of maximization of the system reliability and minimization of system cost has been taken into consideration.

8.2 Review of Literature

In the recent past various researches in India and abroad have made an attempt to evaluate the reliability parameters of several type of complex redundant system by assuming various distributions for the time to failure and repair. Initially Hosford (1960) considered a simple system assuming its failure and repair time distributions to be exponential time distributions. Thus to cater for these type of situations, the various researchers assumed repair time distributions are to be other than exponential, i.e. weibull, gamma, and general time distributions, etc. and evaluated the reliability parameters for simple an complex systems.

To tackle with general distributions, the technique like imbedded Markov chains and phase techniques, etc. have been widely used to solve the problems. A more efficient technique named as "Inclusion of Supplementary Variables" was developed by Cox (1955) [6]. In recent years, various researchers have evaluated the availability/reliability of several complex systems consisting of two or more classes by introducing the concept of redundancy, repair facility with or without priority (Kumar (1992); Kumar (1993); Pan (1997); Philp (1997)) [7–10]. Gupta and Aggarwal (1984) have of course considered cost analysis of three-state repairable systems [11], but that was only in initial stage and needs a further development to meet the challenging requirements of the present day complexities of modern equipment particularly in defence and other private industries.

Besides, both availability analysis and analysis of different complex system reveal that evaluation of system reliability is a basis requirement for all reliability studies. Dhillon (1983), Gupta et al. (2001), Gupta and Agarwal (1984), Pirie and Bendell (1984), Goel and Gupta (1984), have proposed numerous methods to determine symbolic reliability expressions for complex system [12–16].

Tuteja and Malik (1992) have done reliability analysis of a single unit system with three possible modes of the unit: normal, partial failure and complete failure [17].

Gupta et al. (2001) [12] have done analytical study of a complex standby redundant system involving the concept of multi failure.

Singh (2002) introduced copula method in stochastic dependence modeling in environmental hydrology [18]. In his work he derived multivariate frequency distributions.

Melchiori (2003) described various methods for choosing right Archimedean Copula [2].

Bedford (2004) described copulas, degenerate distributions and quintile tests in the complete risk problem [19].

Kumar and Singh (2008) and Kumar et al. (2017a) have applied copula-based method for analyzing the reliability of complex systems [20, 21].

Pant and Singh (2011), Pant et al. (2015a), Kumar et al. (2016) and Kumar et al. (2017b) have applied various optimization techniques for optimizing system reliability and its cost [22–25].

8.3 Material and Methods

In today's technological world nearly all of us depend upon the continued functioning of complex systems and equipment for our day to day activities whether it is related to health safety, mobility or economic welfare. We expect our electronic gadgets, car, computers, etc. to function properly whenever we need them.

8.3.1 Supplementary Variable Technique

It was developed by Cox and by means of introducing some supplementary variables he made the process Markovian.

Consider a system which is complex in nature and in which repair follows general time distribution. It is clear that at a particular time 't', the system will be either in operating mode or in the failure mode. If the system is in the failed mode at time 't', the probability of transition to the operable mode cannot be determined unless the elapsed repair time at that time 't' is specified. A supplementary variable 'x', representing elapsed repair time of the failed unit is introduced and as such is defined as the probability that at time 't', the system is in the failed mode and elapsed repair time lies in the interval $(x, x + \Delta x)$. Thus the process becomes Markovian. It is also interesting to note that such supplementary variable which we introduce in intermediate stage will disappear automatically at the solution stage.

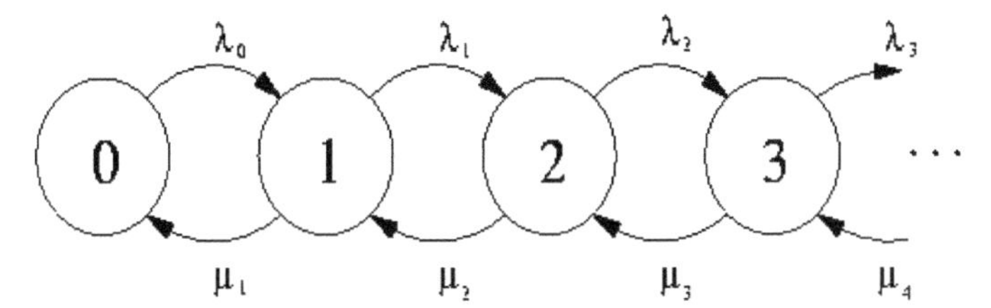

Figure 8.1 Model graph for a birth–death process.

8.3.2 Birth–Death Processes

Birth Death processes have important role in reliability analysis as well as in queueing theory. When analyzing queueing systems useful classes of Markov processes are *birth–death* processes (Figure 8.1).

The state space of the birth–death process is $\{0, 1, 2, 3, ...\}$. The intensity matrix Q which is tri-diagonal type is given by

$$Q = \begin{bmatrix} -\lambda_0 & \lambda_0 & 0 & 0 & 0 & \dots \\ \mu_1 & -(\lambda_1 + \mu_1) & \lambda_1 & 0 & 0 & \dots \\ 0 & \mu_2 & -(\lambda_2 + \mu_2) & \lambda_2 & 0 & \dots \\ . & . & . & . & . & . \\ . & . & . & . & . & . \\ . & . & . & . & . & . \end{bmatrix}$$

intensity matrix. Certain types of queuing systems are suitably modeled by birth–death processes. The numbers $\{\lambda_i\}$ and $\{\mu_i\}$ are interpreted as the arrival rate of the queue and service rate of the server, respectively.

The importance of obtaining highly reliable systems and components has been recognized in recent years. This high degree of reliability is usually achieved by introducing redundancy and service facility. The choice of redundancy depends on the systems configuration. If the failure detection and switching over devices are extremely reliable, the standby redundancy technique must be used in design state as it provides theoretically much higher reliability than parallel redundancy. In standby arrangements, when the main unit in action fails, the identical standby unit is put into the operations through switch.

Redundancy is the provision of alternative means of parallel paths in the system for accomplishing a given track such that all means must fail before causing a system's failure. In fact, the safety of the complex systems can be increased to a greater extent by employing the technique of redundancy.

Redundancy is one of the well-established concepts by which the operating characteristics of a system can be improved.

Anuj et al. (2008) consider two n-unit standby redundant systems designated N and S with repair facility incorporating the concept of imperfect switching under human failure and general repair time distribution. The failure rates of units and repair of the switch are statistically independent.

So far researchers have never taken an important aspect of repairs, i.e. when there are two different types of repairs possible between two transition states, how will this affect the reliability of system as a whole. Anuj et al. (2008) and Anuj et al. (2017) emphasized on this aspect and evaluated the reliability of the system under this consideration. The availability analysis of such type of systems has got their importance in industries, e.g. vehicle.

8.3.3 Multi-objective Particle Swarm Optimization

The objective of computational swarm intelligence is to model the collective behavior of biological populations; it may be the behavior of insect colonies or other animals. The algorithm so developed can be used to solve various optimization problems in general and complex reliability optimization problem in particular. Under this prism, MOPSO is a popular swarm intelligence method which has been used in solving optimization problems. MOPSO is inspired by the ability of flocks of birds, schools of fish and other groups of animals to adapt to their environment and doing their day to day activities like finding food, avoiding the predators by implementing an "information sharing" approaches among themselves, hence, developing an evolutionary advantage. The PSO algorithm, proposed by Eberhart and Kennedy (1995) [26], is used for solving optimization problems [22, 27]. PSO particles update their position by the mean of following two equations. Equation (8.13) calculates a new velocity for each particle based on its previous velocity and Equation (8.14) updates each particle's position in search space.

$$V_{id}^{k+1} = wV_{id}^{k} + c_1 r_1 \left[p_{id}^{k}(t) - x_{id}(t)\right] + c_2 r_2 [p_{g}^{k}(t) - x_{id}^{k}(t)] \qquad (8.13)$$

$$x_{id}^{k+1}(t+1) = x_{id}^{k}(t) + v_{id}^{k+1}(t+1) \qquad (8.14)$$

where k = iteration number; d = 1,2,3……, D; i = 1,2,3…. N; N= swarm size; w = inertia weight, which controls the momentum of particle by weighing the contribution of previous velocity, $c_1$and c_2 are positive constants called acceleration coefficients; $r_1$and $r_2$random numbers uniformly distributed between [0,1]. PSO has also been extended for solving the multi objective problems (MOP), which is generally known as the multi-objective particle swarm optimization (MOPSO) [28]. The MOPSO-CD is swarm-based evolutionary multi objective optimization technique which uses the crowding distance technique [29]. Flow chart of MOPSO-CD is given in Figure 8.2.

8.3.4 Mathematical Model and Reliability Block Diagram

A mathematical model has a number of advantages over the other models, e.g. it may not always be possible to conduct experimentation with scaled down models. The mathematical models and reliability block diagrams can be used to gain new understanding about some phenomenon: e.g. flow of blood in arteries, spread of rumors; to obtain the response behavior of a system: e.g. flight of rocket; to design a complicated piece of equipment: e.g. in space research; design of artificial joint: to optimize some performance such as profits of a company; to make prediction about the system: e.g. climate predictions, etc.

8.3.4.1 Complex bridge system

The source of this problem is Mohan and Shanker (1987) [30]. The block diagram is shown in Figure 8.3.

The system has five components, each having component reliability $r_j\,(j = 1, 2, 3, 4, 5)$.

8.4 Results and Discussion

The overall reliability of the complex bridge system, which is probability of success of system, is given by Tillman et al. (1980) [31].

$$\begin{aligned} R_S = {} & r_1r_4 + r_2r_5 + r_2r_3r_4 + r_1r_3r_5 + 2r_1r_2r_3r_4r_5 - r_1r_2r_4r_5 \\ & -r_1r_2r_3r_4 - r_2r_3r_4r_5 - r_1r_2r_3r_5 - r_1r_3r_4r_5 \end{aligned} \tag{8.15}$$

The cost of jth component is taken as given by Misra (1973):

$$C_S = a_j \exp\left[\frac{b_j}{(1 - r_j)}\right] \quad j = 1, 2, 3, 4, 5 \tag{8.16}$$

Thus the overall system cost is given by:

$$C_S = \sum_{j=1}^{5} a_i \exp\left[\frac{b_j}{(1 - r_j)}\right] \tag{8.17}$$

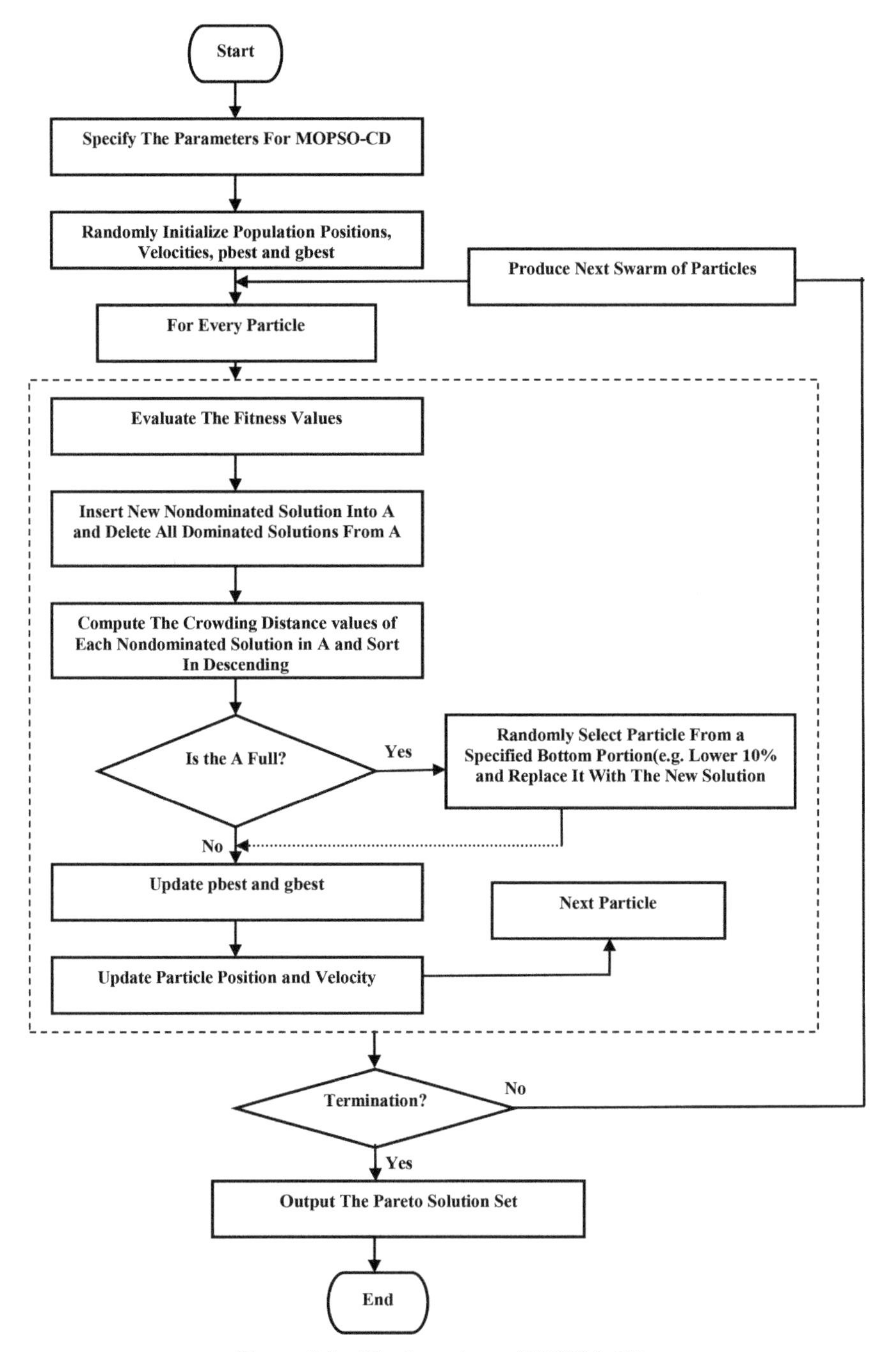

Figure 8.2 The flow chart of MOPSO-CD.

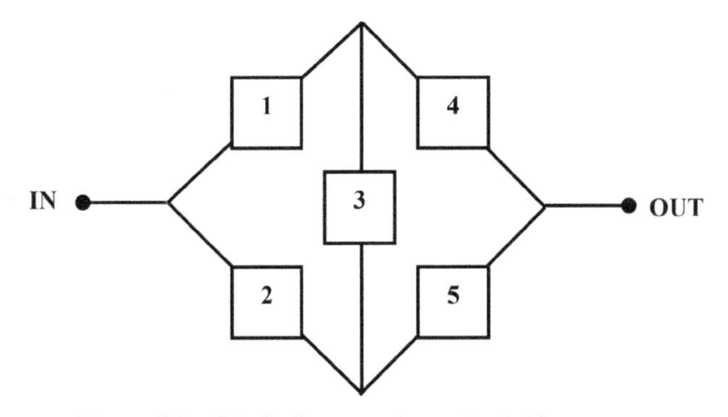

Figure 8.3 Block diagram of complex bridge system.

Table 8.1 Parameters used in MOPSO-CD for complex bridge system

Pop size	Max gen	Mutation prob.	Archive size	C_1	C_2	W
400	800	0.6	200	1	1	0.6

Now the MOOP problem is to determine the reliability of components, which minimizes both system unreliability and system cost. Thus the mathematical formulation of the problem is:

$$\begin{aligned} &\text{Find } (r_1, r_2, r_3, r_4, r_5) \text{ to minimize } (Q_s, C_s) \\ &\text{Subject to} \quad 0 \le r_j \le 1, \quad j = 1, 2, 3, 4, 5 \end{aligned} \tag{8.18}$$

Solving the same problem by MOPSO-CD, the Pareto optimal front obtained is shown in Figure 8.4. Parameters used in MOPSO-CD are given in Table 8.1. This problem is earlier solved by Ravi et al. (2003) [4]. They have solved this problem by fuzzy aggregation method. Their method does not give the entire set of solutions.

A good Pareto optimal front should be as diverse as possible and as close as possible to the true Pareto optimal front. We have tried to find out the parameter settings for obtaining such a Pareto optimal front and one of the best Pareto optimal fronts obtained was shown in Figure 8.4 for complex bridge network.

8.5 Conclusion and Summary

This chapter is very informative in nature and contains history related to reliability with several definitions of reliability given by different authors, definitions of reliability measuring parameters and reliability optimization.

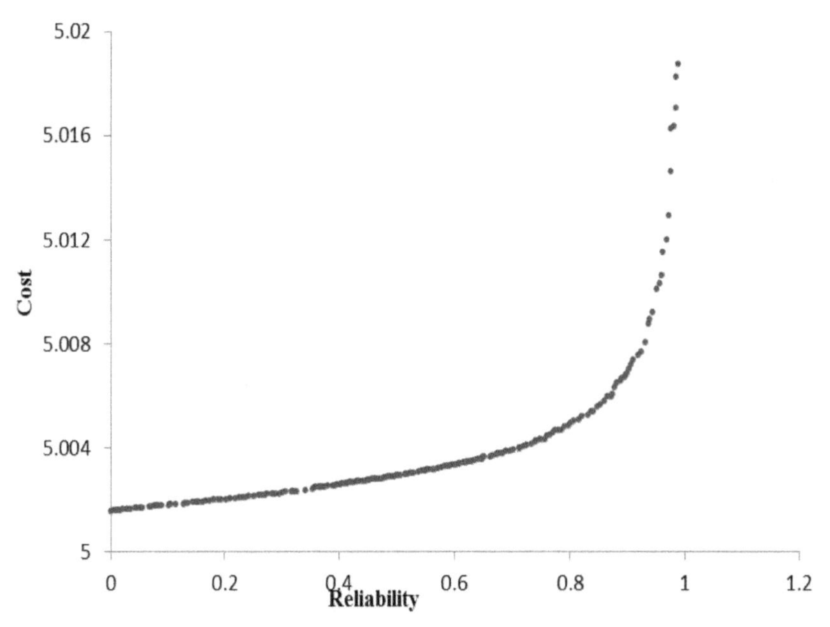

Figure 8.4 Pareto front obtained by MOPSO-CD for complex bridge system.

Review of literature section contains a detailed survey on literature, recent research papers related to reliability theory. Material and methods section discusses about some methods like supplementary variable technique, birth–death processes, multi-objective particle swarm optimization and general model and block diagram which are widely used in numerical computations of reliability, Cost, MTTF, and interpretation of result related to different models. A block diagram of complex bridge system is also presented in the same section. Pareto-optimal front obtained by MOPSO-CD form complex bridge system is given in Figure 8.4 of results and discussion section. The simulation results show that MOPSO-CD is able to generate a well-distributed Pareto optimal set in a single run for a decision maker (DM) to choose from. It gives a DM variety of choices so that he/she could select the most convenient optimal solution according to his/her level of satisfaction in a posterior decision environment.

References

[1] Nelson, R. B. (1999). *An Introduction to Copulas*, 2nd edn. New York, NY: Springer.

[2] Melchiori, M. R. (2003). Which archimedean copula is right one? *Yieldcur. J.* 2003, 1–18.
[3] Ravi, V., Reddy, P. J., and Zimmermann, H. J. (2000). Fuzzy global optimization of complex system reliability, *IEEE Trans. Fuzzy Syst.* 8, 241–248.
[4] Kumar, A., Pant, S., Ram, M. (2016a). System reliability optimization using grey wolf optimizer algorithm. *Qual. Reliab. Eng. Int.*, 366. doi: 10.1002/qre.2107
[5] Pant, S., Kumar, A., and Ram, M. (2017a). Reliability optimization: a particle swarm approach. *Adv. Reliab. Syst. Eng.*, 163–187.
[6] Cox, D. R. (1955). "The analysis of non-Markov stochastic processes by the inclusion of supplementary variables", *Proceedings of the Cambridge Philosophical Society. Mathematical and Physical Sciences*, 51, 433–441.
[7] Kumar, D., Singh, J., and. Pandey, P. C. (1992). Availability of the crystallization system in the ugar industry under common cause failure, *IEEE Trans. Reliab.* 41, 85–91.
[8] Kumar, D., and. Pandey, P. C. (1993). Maintenance planning and resource allocation in urea fertilizer plant, *Qual. Reliab. Eng. Int.* 9, 411–423.
[9] Pan, J. N. (1997). Reliability prediction of imperfect switching system subjected to multiple stresses, *Microelectron. Reliab.* 37, 439–445.
[10] Philp, K.W. and Deans, N. D. (1997). Comparative redundancy, an alternative to triple modular redundant system design, *Microelectron. Reliab.* 37, 581–585.
[11] Gupta, P. P., and Agarwal, S. C. (1984a). A parallel redundant complex system with two types of failure under preemptive-repeat repair discipline, *Microelectron. Reliab.* 24, 395–399.
[12] Dhillon, B. S., and Natesen, J. (1983). Stochastic analysis of outdoor power systems in fluctuating environment. *Microelect. Reliab.* 23, 867–881.
[13] Gupta, P. P., Singh, S. B., and Goel, C. K. (2001). Analytical study of a complex standby redundant system involving the concept of multi failure-human failure under Head-of-Line repair policy, *Bull. Pure Appl. Sci.* 20, 345–351.
[14] Gupta, P. P., and Agarwal, S. C. (1984b). Cost analysis of a 3-state 2-unit repairable system, *Microelectron. Reliab.* 24, 55–59.
[15] Pirie, K. and Bendell, A. (1984). Optimum component switching for system that operates and idle, *Microelectron. Reliab.* 24, 769–780.

[16] Goel, L. R., and Gupta, P. (1984). Stochastic analysis of a two-unit parallel system with partial and catastrophic failures and preventive maintenance, *Microelectron. Reliab.* 24, 881–883.

[17] Tuteja, R. K. and Malik, S. C. (1992). Reliability and profit analysis of two single-unit models with three modes and different repair policies of repairmen who appear and disappear randomly, *Microelectron. Reliab.* 32, 351–356.

[18] Singh, V. P. (2004). Stochastic dependendence modeling in environmental hydrology, *Paper Presented in Sir Mokshagundam Visvesvaraya Memorial Lecture Series,* Pantnagar.

[19] Bedford, T. (2004). Copulas, degenerate distributions and quantile tests in competing risk problems. *Appl. Probabil. Trust*, 8.

[20] Kumar, A., and Singh, S. B. (2008). Reliability analysis an n-unit parallel standby system under imperfect switching using copula. *Comput. Model. New Technol.* 12, 47–55.

[21] Kumar, A., Pant, S., and Singh, S. B. (2017a). Availability and cost analysis of an engineering system involving subsystems in series configuration. *Int. J. Qual. Reliab. Manag.* 34.

[22] Pant, S., and Singh, S. B. (2011). "Particle swarm optimization to reliability optimization in complex system," *IEEE International Conference on Quality and Reliability*, Bangkok, 211–215.

[23] Pant, S., Anand, D., Kishor, A., and Singh, S. B. (2015a). A particle swarm algorithm for optimization of complex system reliability. *Int. J. Perform. Eng.* 11, 33–42.

[24] Kumar, A., Pant, S., Singh, S. B. (2016b). Reliability optimization of complex system by using cuckoos search algorithm, *Math. Concept Appl. Mech. Eng. Mech.*, 94–110.

[25] Kumar, A., Pant, S., Ram, M., and Singh, S. B. (2017b). On solving complex reliability optimization problem using multi-objective particle swarm optimization, *Math. Appl. Eng.*, 115–131.

[26] Eberhart, R., and Kennedy, J. (1995). A new optimizer using particle swarm theory. *In Proceeding of International Symposium on Micro Machine and Human Science*, Nagoya, 39–43.

[27] Pant, S., Kumar, A., Kishor, A., Anand, D., and Singh, S. B. (2015b). "Application of a multi-objective particle swarm optimization technique to solve reliability optimization problem," in *Proceeding of IEEE International Conference on Next Generation Computing Technologies*, Dehradun, 1004–1007.

[28] Coello, C. A. C., Pulido, G. T. and Lechuga, M. S. (2004). Handling multiple objectives with particle swarm optimization. *IEEE Trans. Evol. Comput.* 8, 256–279.

[29] Raquel, C. R. and Naval, P. C. (2005). An effective use of crowding distance in particle swarm optimization. *Genet. Evol. Comput. Conf.*, 257–264.

[30] Mohan, C., and Shanker, K. (1988). Reliability optimization of complex systems using random search technique. *Microelect. Reliab.* 28, 513–518.

[31] Tillman, F. A., Hwang, C. L. and Kuo, W. (1980). *Optimization of Systems Reliability*. New York, NY: Marcel Dekker Inc.

Pant, S., Kumar, A., Singh, S. B., and Ram, M. (2017b). A modified particle swarm optimization algorithm for nonlinear optimization. *Nonlinear Stud.* 24, 127–138.

Pant, S., Kumar, A., and Ram, M. (2017c). Flower pollination algorithm development: a state of art review. *Int. J. Syst Assur. Eng. Manag.*, 1–9. doi: 10.1007/s13198-017-0623-7.

9

Tree Growth Models in Forest Ecosystem Modeling – A Tool for Development of Tree Ring Width Chronology and Climate Reconstruction

Rajesh Joshi and Rupesh Dhyani

G. B. Pant National Institute of Himalayan Environment and Sustainable Development, Kosi-Katarmal, Almora, India

Abstract

The modeling techniques have become an integral part of ecological research particularly for reducing various complexities and inconsistencies for designing field studies and estimate experimental knowledge. Forest ecosystem models are designed to detect, quantify and reproduce the dynamics within the forest ecosystem. These models are vital tools to extrapolate forest ecosystem processes at different spatial and temporal scales such as from the point to the landscape level or from the landscape to the plot level. While applying modeling techniques to ecological system, various complexities such as scale, and long term data availability, remain as persistent issues which need to be addressed to ascertain a reasonably stable modeling method. Although, the complexity of the different approaches is quite diverse, each method adequately addresses the distinct objectives and to achieve the expected output. Therefore, choosing the correct level of model complexity becomes an important step within the modeling process because it ensures optimized use of available resources, and the reliability of the model output. Taking all such issues in consideration, this chapter explains the notion of ecosystem modeling through a system analytical approach in designing an ecosystem model. Further, the concepts of different forest ecosystem models (e.g. growth models, succession models, mechanistic models and hybrid models) together with their application potential are discussed here. Furthermore, in this chapter, the application of tree growth models along with its various components is explained and demonstrated with the help of an example from a case study of development of tree ring chronology for *Pinus roxburghii* (commonly known as chir pine or longleaf Indian pine) from Pithoragarh region of Uttarakhand state in India. The developed chronology is useful in reconstruction of past climate. A major obstruction to the application of process-based models is the operational implementation of these models. Therefore, an integrated approach of research in ecological modeling is imperative for decision-making in forest management.

9.1 Introduction

9.1.1 Notion of Ecosystem Modeling

Forests ecosystems have been severely affected by various anthropogenic activities and natural phenomena. Globally, these activities have resulted in reduction in forest-covered area affecting the patterns of various tree-species, management of forest resources and timber production adversely. Forest ecosystem modeling is one of the tools used to assess and understand these changes. Since field-based primary observations generally exist for small duration restricted to only a few sites; mathematical models become very vital for spatial and temporal extrapolation of ecological and environmental changes [1]. However, while simulating various processes and estimating changes, various inconsistent issues (e.g. minimalism, data dependency and practicality) must be addressed to ascertain a reasonably stable modeling method. In some of the research fields, advancement in scientific understanding is crucial which is often considered as limitation in modeling process. In such circumstances, empirical/statistical methods can help in selection of suitable research design and methods of analysis for data generated from the study. In case of missing information, ordinary empirical/statistical methods can be used to establish relationship between input and output parameters after proper gap filling to develop a comprehensive perceptive of the process.

In research, mathematical models are defined as a simplified mathematical illustration of a physical processes comprising of a certain level of generalization which is much simpler and capable of explaining the processes of real-world through a prototype [2]. Such models exist in a multiplicity of structure which range from theoretical to computer-based simulation models [3]. A system-based logical approach is required for building a

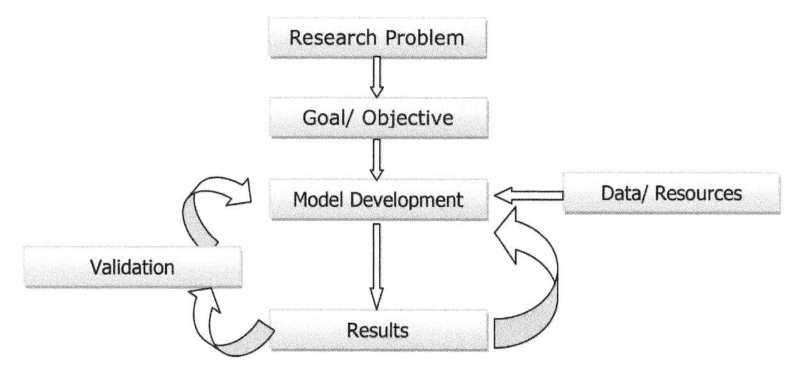

Figure 9.1 System analytical approach in the design of ecosystem model.

model which includes (i) formulation of problem, (ii) resource planning, (iii) research design and data collection, (iv) model development and its calibration, (v) validation of the model and (vi) sensitivity analysis (Figure 9.1). To follow the system analytical approach, it is very crucial that parameters, constraints and objective functions should be defined as per the criterion of the study so that various inconsistency issues are addressed. While defining the objectives of study, functionality and structure/layout of the model is defined. Further, considering the scaling in space and time, often modeling approaches are constrained by non-availability of required dataset. Therefore, planning is required for optimization of resources so that the cost and time involved in data collection could be minimized. In case the models applied on a higher temporal and spatial scale, the dataset used for initialization and calibration of the model has key impact on the reliability of the model output. Generally, models are developed based on the available knowledge; hence validation and performance evaluation of the model is carried outside the calibration data frame to ensure reliability of the model and assess the error in the model. While using ecosystem models, it is imperative to know how a model responds to original/initial conditions as well as under different changed scenario. These kinds of simulations under changed scenario are vital for sensitivity analysis of the modeling process which is a foremost step in assessing and testing the reliability of model output.

Different procedures for data collection and integration are required for different ecosystem processes acting at diverse levels as some may act on an individual level (such as energy budget) whereas other processes may act at a larger scale such as regional or global scale (e.g. climate change). Similarly, some processes operate within a very short span of time (e.g. gas exchange

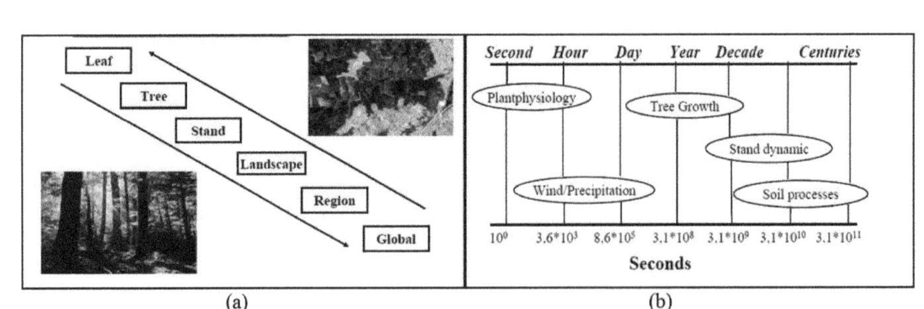

Figure 9.2 Spatial and temporal scaling issues (a) in space and time (b) in time within ecosystem modeling (Adopted from Hasenaurer et al., 2000).

within the stomata), while others take considerably very long time years to have an impact on a forest ecosystem (e.g. geological formation).

Accommodating these differences in scale is critical for combining different processes. Therefore, to ensure a "well-balanced" modeling approach, three modeling criteria (i.e. simplicity, observability and biological practicality) are needed to be addressed so that all the model variables are consistently and equally incorporated. However, if biological practicality is not correctly addressed, an imbalanced approach may lead to ambiguous results. Models are developed to integrate knowledge, explain relations between eco-physiological procedures, formulate hypothesis, identify gaps and to develop decision support system to execute scenario-based analysis. With the enhancement of knowledge on science and technology, different mathematical models have been developed to analyse change in forest ecosystem. However, from a conceptual point of view, forest ecosystem model concepts can be categorized in three: (i) growth models, (ii) succession models and (iii) mechanistic models [4].

9.1.1.1 Growth and yield models

The periodical forest inventories are used to assess the conditions of forest. Forest resource maps integrated with model-based growth and predictions offer key knowledge and information for forest planning and management. Conventionally, forest management rely on yielding tables which assume that the forest stands are pure, even-aged and undergo only one treatment [5]. Based on which, for a given site characteristics, mean stand development over time can be described. The main aim of growth and yield model, which exists more than 125 years, is to predict the future stand development for management of forest resources and development of sustainable management

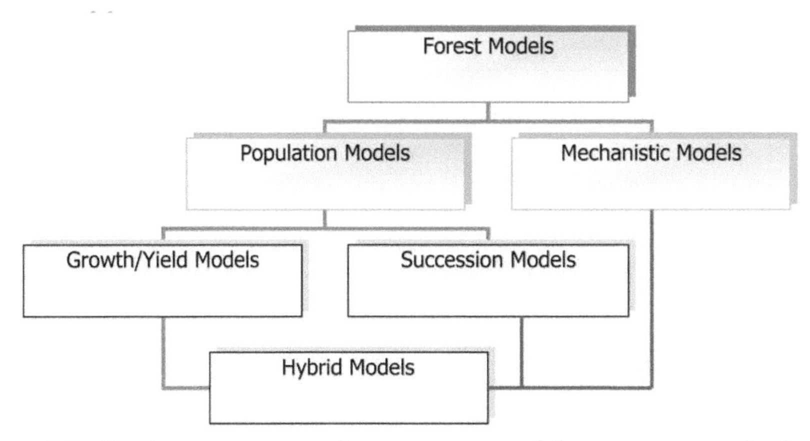

Figure 9.3 The three most-common forest ecosystem modeling concepts as per implementation principles.

plans. With the change of paradigm in management of forest resources, initial yield tables have been improved and to some extent substituted by tree growth models to improve reliability of the model output. As a result, tree growth model, operational at the individual tree level developed [6, 7]. The key benefit of these models is that they are independent from any specific mixture, age and stand treatment. Some of the common applications of growth and yield models are predictions of periodical growth rates (diameter and height increment) and survival probability to ensure sustainable forest management [4, 8].

9.1.1.2 Succession models

In a forest ecosystem old trees are succeeded by new trees growing in forest gaps or patches. For modeling such processes Succession, Gap, or Patch models are generally used to describe the reproduction growth and mortality of trees [9]. The foremost succession model (JABOWA I) was developed by Botkin et al. in 1972 [10] for describing the stand dynamics which was followed by several developments and application of succession models [1, 2, 11, 12]. In succession models, predefined species-specific growth potential is calculated by simulating the management models' growth potential of different sites. These models are explicitly used to assess the impact of different parameters (such as temperature, water and nutrients) on growth of a tree by applying multipliers, whereas in case of the management models individual tree-based competition procedures are used to predict the periodical growth rates of trees. Although succession model simulates tree growth, yet they are mainly used to describe vegetation patterns over different time scales. Functioning of the succession model is further enhanced while analyzing the interaction between the key factors of growth (e.g. energy, temperature, water, nutrients). Nowadays, these models are used to assess the vegetation pattern and distribution under future climate change.

9.1.1.3 Biogeochemical-mechanistic models

Mechanistic models, considered analytical tools to understand ecosystem behavior, describe the energy exchange within and through a forest ecosystem. The foremost aim of mechanistic models is to analyze interactions between plants and their surrounding environment. Therefore, they are often termed as process or biogeochemical models (BC-Models). Unlike tree population models (e.g. management models, succession models), these models integrate a mechanistic explanation of the interactions between plants and surrounding environment [13]. BGC models operate on a finer temporal resolution (i.e. monthly to daily) to simulate the carbon, water and nitrogen cycles for generalized biome or species [14] (Thornton 1998) types. BGC models can also operate on a smaller scale such as stand level to simulate net primary productivity (NPP) (Farquhar et al. 1980). BGC models need data on meteorological parameters (e.g. daily minimum and maximum temperature, incident solar radiation, vapor pressure deficit, precipitation) as model input variables. In case of non-availability of values for these parameters, the same may be interpolation from a similar site within the area using mathematical algorithms [15]. The specific applications of BGC models include modeling the carbon cycle, evaluation of response of tree growth to environmental changes [16] and explanation of the water and nutrient cycle [17, 18]. Improvements in BGC modeling techniques are primarily required for augmentation of the species representation and inclusion of management scenarios [18, 19].

9.1.1.4 Hybrid models

The integration of different concepts is important to further strengthen the knowledge related to application of modeling for simulation of the ecosystem processes. Integration of diverse concepts and theories leads to inconsistent results. A typical example is the implementation of forest management wherein tree growth models are explicitly designed to be sensitive to density, in such a case, large scale BGC models tend to overestimate. This is because of the reason that density of the stand changes within forest; this may often lead to an overestimation of forest growth response to forest management if time lag effects are not considered [20]. Thus BGC-models may tend to overestimate NPP following thinning of the stand. Other examples are: (i) typical gap models assume a constant height/diameter ratio [9] and (ii) Growth and Yield model assumes constant site conditions to forecast timber growth. Therefore, to overcome the principle conceptual differences of the modeling concepts, various techniques are integrated to develop the fusion model called "Hybrid models" so as to improve the applicability of modeling tools.

9.2 Tree Growth Models

Tree growth rate fluctuates within the entire growing span of a tree. At the different stages of tree life, the growth rate of tree increases or decreases monotonically. Variation in growth trend can be captured by observing the growth trend of tree and fitting growth models to the time series data of tree growth. Methods of eliminating the trends (i.e. the biological age-related trend) in tree-ring measurements are divided into three major categories, (i) deterministic curves, (ii) stochastic curves and (iii) empirical curves [21]. These curves help in understanding the interaction between plant growth/development with environmental factors including temperature, rainfall, water availability and soil characteristics. To model these interactions we need to take into account the physiological processes involved in growth. Growth of tree is described in the form of tree ring which represents the age of a tree (e.g. a complete ring represents one year growth of tree). As the age increases many environmental factors affect the growth of tree; the environmental effects on tree growth are reflected by fluctuation in growth curve of tree. These fluctuations give rise to a tree ring chronology or time series of tree ring width. To estimate the effect of environmental factors on tree growth, non-environmental factors and noise are removed and appropriate growth curve is fitted on time series.

9.2.1 General Linear Aggregate Model

Let us assume a single tree-ring series with a uniform radius as a linear aggregate of several sub-series. In such a case, the ring width is represented as the sum of environmental factors

$$RW = C + A + D_1 + D_2 + E \quad (9.1)$$

where, RW – the well-dated tree-ring widths measured along uniform radius, C – the climatically-related growth variations common to a stand of trees including the mean persistence of these variations due to physiological preconditioning and interaction of climate with site factors, A – the age-related growth trend, D_1 – the endogenous disturbance pulse originating from forces within the forest community, D_2 – the exogenous disturbance pulse originating from forces outside the forest community and E – the series of more or less random variations representing growth influencing factors unique to each tree or radius within the tree. This equation explains one year functional relationship of tree ring growth with several factors [22]. The assumption of linearity and implicit independence between the sub-series is necessary for simplification. The purpose of this model is not to describe exact relationships between the subseries. Rather, it allows argument on certain properties of each component separately from the others as a necessary step in developing a standardization method that models the nature of the tree-ring series more adequately.

9.2.2 Growth Curve for Detrending Tree Growth Time Series

The main aim of standardizing a tree growth time series is to remove non-climatic signal (i.e. A, Dl, D2 and E) and filter the climate signal (C) which is further used for paleo-climatic studies. The process of modeling by which non-climatic factors are removed from tree growth time series is called standardization [22]. After estimation of tree growth curve, the standardized tree growth indices are calculated as

$$RW_t = \frac{r_t}{g_t} \tag{9.2}$$

where RW_t is the index, r_t is observed tree growth and g_t is estimated tree growth for year *t*. After standardization, the resulting tree growth time series reduces to dimensionless indices. To fit a growth curve on tree growth time series various assumptions are to be made; these assumptions characterize the environmental conditions under which tree grows. The deterministic curves, stochastic curves and empirical curves are used for detrending the tree growth time series.

9.3 Deterministic Curves

9.3.1 Negative Exponential Curve

If the nature of tree growth time series is decreasing with the age, then negative exponential curve is fitted to tree ring series. The negative exponential curve as a model is expressed as

$$y = ae^{-bx} \tag{9.3}$$

or

$$\log y = \log a - bx \text{ in linear form,}$$
$$\text{y} \rightarrow 0 \text{ as x} \rightarrow \infty$$

This model is not suitable for very old conifer trees because ageing trend of these kinds of trees frequently approaches to a constant level of growth. In view of this, the modified exponential growth equation is used in place of Equation (9.3). The modified exponential curve is expressed as:

$$w_t = ae^{-bx} + k \tag{9.4}$$

where w_t is ring width for year *t*, *a* is ring-width at year zero (if *k* is negligible), *b* is the slope of the decrease in ring width and *k* is the minimum ring width, which is asymptotically approximated for large values of *t*. The modified curve is considered as more flexible for standardization of tree ring data for open canopy environment. This is basic assumption to fit the negative exponential curve to tree growth time series. Based on data, the modified curve is an excellent age trend model. The main limitation of this curve is that it cannot model the juvenile growth period of increasing growth rate [23].

9.3.2 Linear Regression Curve

In case, if the estimated value of either *a* or *b* is negative, a linear regression model is fitted to the data usually with slope ≤ 0 [23, 24]. Therefore, in such a case, instead of fitting a modified negative exponential curve, regression line could be selected mainly because of numerical instabilities. The linear regression model is expressed as

$$y = a + bx \tag{9.5}$$

If the priori biological model for estimating the age trend is required, the slope coefficient *b* may be constrained to negative or zero.

9.3.3 Hugershoff Growth Curve

The accelerated growth of tree rings, particularly close to the pith, cannot be simulated by the monotonic increasing or decreasing curves of the linear or the negative exponential function. In such a case, Hugershoff curve is fitted on the tree ring time series data [25] as follows:

$$f(t) = at^{b}e^{-ct} + d \tag{9.6}$$

where t represents age, $f(t)$ represents the ring-width value for t, the superscript b denotes the power, c indicates the exponential argument and d is the intercept of the mode. If b equals 0, Equation (9.6) becomes the exponential curve. Similarly, the Hugershoff curve becomes a linear function when b and c equal 1 and 0, respectively. The Hugershoff curve is considered as more representative than the other deterministic curves.

9.4 Stochastic Curves

9.4.1 The Smoothing Spline Curve

The smoothing spline curve is used to remove the non-climatic variance of tree ring width data series [26]. In this method, no assumption is made about the nature of growth curve to be used for standardization. The spline curve can range continuously from a linear least square fit to cubic interpolation; it is therefore far more flexible than polynomial function and provides a more likely fit to the data series. The smoothing spline algorithm, as proposed by Reinsch (1967), minimizes the total squared curvature of the spline and is defined as follows [27]:

$$\int_{x_0}^{x} [g''(x)]^2 dx \tag{9.7}$$

under the constraint

$$\sum_{i=0}^{n} \left[\frac{g(x_i - y_i)}{\delta y_i} \right]^2 \leq S \tag{9.8}$$

where y_i is the input data series, δy_i is the series of weights and S is the scaling parameter. The quantities δy_i control the extent of smoothing and are implicitly rescaled by varying S. The standardization curve should passes through the local average of ring width series. In this case, δy_i is set to zero or all i. The above expression then reduces to an un-weighted residual sum of squares. Smoothing spline results in highly improved standardized ring width series in comparison to straight line or negative exponential curve.

9.4.2 Friedman's Super Smoother

Friedman (1984) described a stochastic function Friedman variable smoother for modeling biological growth trend in a tree ring data [28]. This smoother is applied to a set of bivariate data (x_1, y_1), (x_2, y_2), ..., (x_n, y_n) that produces the function as follows

$$y_i = s(x_i) + r_i, i = 1 \dots n \quad (9.9)$$

where s is a smooth function and r_i are residuals.

In general, (x_1,y_1), $(x_2,y_2)\dots(x_n,y_n)$ are the random samples from a joint probability distribution $P(X, Y)$ and defined as optimal function f for predicting Y as a function of X that minimizes the expected squared difference between Y and $f(X)$.

$$\text{i.e.} \quad E_{X,Y}[Y - F(X)]^2 = \min E_{X,Y}[Y - F(X)] \quad (9.10)$$

For estimating conditional expectation of $\underline{Y}$ smoothers procedure can be used. This implies that for estimating $f(x) = E[Y|X = x]$ we have to find

$$Y = f(x) + \varepsilon \quad (9.11)$$

where $f(X)$ is a smooth function and e is an independent and identically distributed random variable with zero expectation.

9.5 Empirical Curves

9.5.1 Regional Curve Standardization Method

Becker, in 1989, described the Curve Standardization technique [29] which was later named as RCS method [30]. The RCS method is useful for the dominant radial growth series. In a tree ring series, the age i of tree is denoted by a_i, where $a_i \in [1, a_{\max}]$ and $a_{\max}$ is the maximum age. The number of rings available at age a_i and date t_j is denoted by n_{ij}. Ring widths at age a_i and year t_j are denoted by $RW_0(a_i, t_j)$ and are further indexed by $k \in [1, n_{ij}]$ when $n_{ij} > 1$.

The RC is calculated for all a_i as:

$$\overline{RW}_{0(a_i)} = \frac{\sum_{j=t_{min}}^{t_{max}} \sum_{k=1}^{n_{i,j}} RW_{0,k}(a_i t_j)}{\sum_{j=t_{min}}^{t_{max}} n_{ij}} \quad (9.12)$$

The RC is smoothed (denoted by RW_0) using a polynomial function fitted by ordinary least squares. A polynomial of degree 4 is selected from successive F-tests for nested model comparison [31]. Each ring $RW_0(a_i, t_j)$ is then standardised to a growth index $I_0(a_i, t_j)$ using the Equation

$$I_0(a_i, t_j) = \frac{RW_0(a_i, t_j)}{\overline{RW}_0(a_i)} \quad (9.13)$$

The regional chronology was calculated as the average of growth indices at each calendar year:

$$\bar{I}_0(t_j)_{size} = \frac{\sum_{i=Ro_{min}}^{Ro} \sum_{k=1}^{n_{i,j}} I_{0,k}(R_0, t_j)}{\sum_{i=Ro_{min}}^{Ro} n_{ij}} \tag{9.14}$$

To facilitate the comparison between the different methods, the annually resolved RCS chronologies are smoothed using a non-parametric cubic spline method. Finally, to express the evolution in growth relative to the same reference year, the RCS chronologies are adjusted to apply cubic-spline smoothers.

9.6 Application of Tree Ring Growth Models – An Example from A Case Study

Application of tree ring growth models is demonstrated with a real world problem. Tree growth model is used to develop the tree ring width chronology which captures the climate tree growth relationship. Tree ring-based study of *Pinus roxburghii* (commonly known as chir pine or longleaf Indian pine) is carried out in Pithoragarh district of Uttarakhand state in India. Approximately 100 tree ring cores were collected from different sites and tree ring

a. Collection of tree core samples | b. Extraction of samples | c. Mounting of samples

d. Polishing of samples | e. Samples (cores) ready for analysis | f. Measurement of tree ring width

chronology was prepared after measurement of tree ring width (plate a–f). The developed chronology was used to reconstruct the past climate for the region.

A negative exponential growth curve is fitted to develop the tree ring width chronology from collected samples. The standardization tree ring width chronology is developed using expected growth curve (which depends on the nature of tree growth curve) as follows:

$$Y_t = a * \exp(-b * t + d) \tag{9.15}$$

and the Ring Width Index is developed using the following equation:

$$RI_t = \frac{R_t}{Y_t} \tag{9.16}$$

where, RI_t – Index value for the year 't'

R_t – Aggregate growth

Y_t – Expected yearly growth

Using Equations (9.15) and (9.16), a raw and standardized chronology is developed as depicted in Figures 9.4 and 9.5.

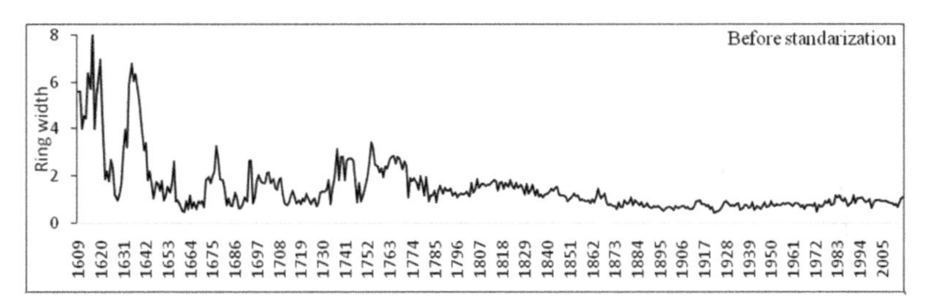

Figure 9.4 The negative exponential growth of a tree with the increasing age.

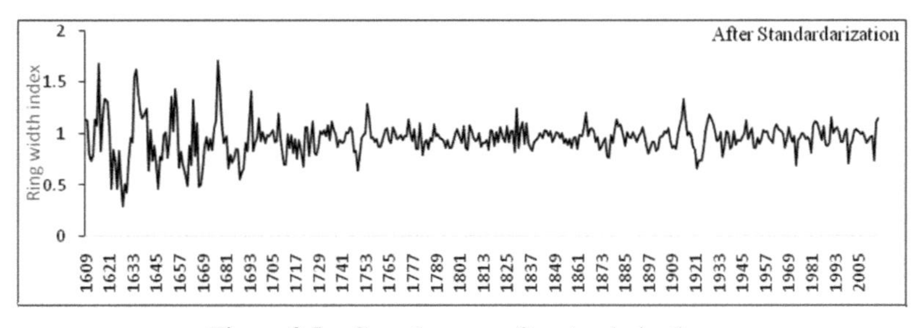

Figure 9.5 Growth curve after standarization.

The chronology thus developed is used to study the climatic control on tree-ring growth of *Pinus roxburghii* in western Himalayan region.

9.7 Conclusion

Forest ecosystem models are designed to detect, quantify and reproduce the dynamics within the forests. These models are vital tools used to extrapolate forest ecosystem processes at different spatial and temporal scales such as from the point to the landscape level (bottom–up approach), or from the landscape to the plot level (top–down approach). Accommodation of these scales is important to ensure a successful application. Forest Growth and Yield Models are considered as the classical approaches to assess forest ecosystem processes. They are designed to provide information for forest management purposes. Some of the common application areas are: (i) understanding the tree population dynamic as it has evolved over time, (ii) estimation of growth rate of trees, (iii) predicting the stand growth, (iv) forest stand regeneration, growth and harvest simulation, (v) climatic control on tree-ring growth, etc. Similarly, a key area of application for succession models is the assessment of potential forest-vegetation patterns. Whereas, mechanistic models are explicitly designed to understand the complex interaction between energy, water and nutrient cycles within forest ecosystems. They are used to assess the interactions and flows within forest ecosystems and the surrounding environment.

Although, the complexity of the different approaches is quite diverse, each method adequately addresses the distinct objectives and to achieve expected output. Therefore, choosing the correct level of model complexity is an important step within the modeling process because it guarantees optimized use of available resources, and the reliability of the model output. Integrating modeling concepts to develop a hybrid model is nowadays appealing and mostly becomes useful. However, the key difficulty in developing a hybrid model is to include conceptual constraints to ensure a reliable output. Further, adding more input variables may substantially increase the random error component of the resulting predictions. Therefore, caution must be taken to develop an integrated hybrid model. Furthermore, tree growth models are based on some assumptions about the process being modelled. Most of the models are used to give some insights into the forest ecosystem processes and help in making rational decisions for management. These models are also used to extrapolate beyond the data which may include some error. Hence, it is very important to exercise caution while interpreting model predictions.

Recent developments in the application of process-based modeling to the management of forest growth and change have integrated the development of evaluation and parameter estimation methods suitable for models with causal structure, and the increasing availability of data applicable to the evaluation of such models. It is obvious that many questions need further investigation; however the quantification of models for any particular site cannot easily be carried out without calibration against system-level data. A major obstruction to the application of process-based models is the operational implementation of these models. The decision-making and analysis in forest management are increasingly being done following more general methods and causal-oriented approach. Therefore, there is need for close interaction between modelers and forestry practitioners for mutual appreciation of methods and approaches used for ecological modeling.

Acknowledgements

This book chapter is an outcome of the study carried out under the National Mission for Sustaining the Himalayan Ecosystem (NMSHE), Task Force 3: Forest Resources and Plant Biodiversity (NMSHE, TF-3), funded by the Department of Science and Technology (DST), Government of India. The funding support received from DST, GoI is duly acknowledged. Further, the authors would like to thank Director, GBPNIHESD, Kosi-Katarmal, Almora for providing necessary facilities to conduct the study.

References

[1] Pacala, S. W., Canham, C. D., Saponara, J., et al. (1996). Forest models defined by field measuremensts: estimation, error analysis and dynamics. *Ecol. Monogr.* 66, 1–43.

[2] Shugart, H. H. (1998). *Terrestrial Ecosystems in Changing Environments*. Cambridge: Cambridge University Press, 537.

[3] Kimmins, J. P. (1987). *Forest Ecology: Models and the Role in Ecology and Resource Management*. New York, NY: Macmillan Publishing Company, 460–474.

[4] Hasenauer, H. (2005). *Sustainable Forest Management: Growth Models for Europe*. Berlin: Springer, 398.

[5] Weise, W. (1880). *Ertragstafeln für die Kiefer*. Berlin: Verlag Springer, 156.

[6] Monserud, R. A. (1975). *Methodology for Simulating Wisconsin Northern Hardwood Stand Dynamics*. Ph.D. thesis, University of Wisconsin, Madison, WI, 156.

[7] Wykoff, W. R., Crookston, N. L., and Stage, A. R. (1982). *User's Guide to the Stand Prognosis Model*. Washington, DC: USDA Forest Service, 112.

[8] Hasenauer, H. (1997). Dimensional relationships of open-grown trees in Austria. *For. Ecol. Manage.* 96, 197–206.

[9] Botkin, D. B. (1993). *Forest Dynamics: An Ecological Model*. Oxford: Oxford University Press, 309.

[10] Botkin, D. B., Janak, J. F., and Wallis, J. R. (1972). Some ecological consequences of a computer model of forest growth. *J. Ecol.* 60, 849–872.

[11] Lexer, M. J., and Hönninger, K. (2001). A modified 3D-patch model for spatially explicit simulation of vegetation composition in heterogenous landscapes. *J. Ecol.* 60, 849–872.

[12] Bugmann, H. (2001). A review of forest gap models. *Clim. Change* 51, 259–305.

[13] Waring, R. H., and Running, S. W. (1998). *Forest Ecosystems: Analysis at Multiple Scales,* 2nd Edn. San Diego, CA: Academic Press, 370.

[14] Pietsch, S. A., Hasenauer, H., Thornton, P. E. (2005). BGC-Model parameters for tree species growing in central European forests. *For. Ecol. Manag.* 211, 264–295.

[15] Hasenauer, H., Merganicova, K., Petritsch, R., et al. (2003). Validating daily climate interpolations over complex terrain in Austria. *Agric. For. Meteorol.* 119, 87–107.

[16] Hasenauer, H., Nemani, R. R., Schadauer, K., and Running, S. W. (1999). Forest growth response to changing climate between 1961 and 1990 in Austria. *For. Ecol. Manage.* 122, 209–219.

[17] Running, S. W., and Coughlan, J. C. (1988). A general model of forest ecosystem processes for regional applications. I. Hydrologic balance, canopy gas exchange and primary production processes. *Ecol. Model.* 42, 125–154.

[18] Pietsch, S. A., Hasenauer, H., Kucera, J., and Cermak, J. (2003). Modelling the effects of hydrological changes on the carbon and nitrogen balance of oak in floodplains. *Tree Physiol.* 23, 735–746.

[19] Pietsch, S. A., and Hasenauer, H. (2001). Using mechanistic modelling with in forest ecosystem restoration. *For. Ecol. Manage.* 159, 111–131.

[20] Petritsch, R., Hasenauer, H., and Pietsch, S. A. (2007). Incorporating forest growth response to tinning within BIOME-BGC. *For. Ecol. Manage.* 242, 324–336.

[21] Cook, E. R., and Kairiukstis, L. A. (eds). (1990). *Methods of Dendro chronology: Applications in the Environmental Science*. Dordrecht: Kluwer Academic Publishers, 394.

[22] Fritts, H. C. (1976). *Tree Rings and Climate*. New York, NY: Academic Press.

[23] Fritts, H. C., Mosimann, J. E., and Bottorff, C. P. (1969). A revised computer program for standardizing tree-ring series. *Tree Ring Bull.* 29, 15–20.

[24] Cook, E. R., and Holmes, R. L. (1986). *Guide for Computer Program ARSTAN*. Tucson, AZ: University of Arizona, 65.

[25] Warren, W. G. (1980). On removing the growth trend from dendrochronological data. *Tree Ring Bull.* 40, 35–44.

[26] Cook, E. R., and Peters, K. (1981). The smoothing spline: A new approach to standardizing forest interior tree-ring width series for dendroclimatic studies. *Tree Ring Bull.* 41,45–53.

[27] Reinsch, C. H. (1967). Smoothing by spline functions. *Numerischemathematik* 10.3, 177–183.

[28] Friedman, J. H. (1984). *Laboratory for Computational Statistics,* Vol. 5. Stanford University Technical Report.

[29] Becker, M. (1989). The role of climate on present and past vitality of silver fir forests in the Vosges mountains of northeastern France. *Can. J. For. Res.* 19, 1110–1117.

[30] Briffa, K. R., Jones, P. D., Bartholin, T. S., Eckstein, D., Schweingruber, F. H., Karlen, W., et al. (1992). Fennoscandian summers from AD 500 – temperature changes on short and long time scales. *Clim. Dyn.* 7, 111–119.

[31] Seber, G. A. F., and Lee, A. J. (2003). *Linear Regression Analysis*, 2nd Edn. Hoboken, NJ: Wiley, 653.

Index

About the Editors

Mangey Ram received the Ph.D. degree major in Mathematics and minor in Computer Science from G. B. Pant University of Agriculture and Technology, Pantnagar, India, in 2008. He has around ten years teaching and research experience after post Ph.D. and has taught several core courses in pure and applied mathematics at undergraduate, postgraduate, and doctorate levels. He is currently a Professor at Graphic Era University, Dehradun, India. Before joining the Graphic Era University, he was a Deputy Manager (Probationary Officer) with Syndicate Bank for a short period. He is Editor-in-Chief of International Journal of Mathematical, Engineering and Management Sciences; & Member of the editorial board of many journals. He is a regular Reviewer for international journals, including IEEE, Elsevier, Springer, Emerald, John Wiley, Taylor & Francis and many other publishers. He has published 125 research publications in IEEE, Springer, Emerald, World Scientific and many other national and international journals of repute and also presented his works at national and international conferences. His fields of research are reliability theory and applied mathematics. Dr. Ram is a senior member of the IEEE, member of Operational Research Society of India, Society for Reliability Engineering, Quality and Operations Management in India, International Association of Engineers in Hong Kong, and Emerald Literati Network in the U.K. He has been a member of the organizing committee of a number of international and national conferences, seminars, and workshops. He has been conferred with "Young Scientist Award" by the Uttarakhand State Council for Science and Technology, Dehradun, in 2009. He has been awarded the "Best Faculty Award" in 2011 and recently Research Excellence Award in 2015 for his significant contribution in academics and research at Graphic Era.

J. Paulo Davim received the Ph.D. degree in Mechanical Engineering in 1997, the M.Sc. degree in Mechanical Engineering (materials and manufacturing processes) in 1991, the Mechanical Engineer degree (MEng-5 years) in 1986, from the University of Porto (FEUP), the Aggregate title (Full

Habilitation) from the University of Coimbra in 2005 and the D.Sc. from London Metropolitan University in 2013. He is Eur Ing by FEANI-Brussels and Senior Chartered Engineer by the Portuguese Institution of Engineers with a MBA and Specialist title in Engineering and Industrial Management. Currently, he is Professor at the Department of Mechanical Engineering of the University of Aveiro, Portugal. He has more than 30 years of teaching and research experience in Manufacturing, Materials and Mechanical Engineering with special emphasis in Machining & Tribology. He has also interest in Management & Industrial Engineering and Higher Education for Sustainability & Engineering Education. He has guided large numbers of postdoc, Ph.D. and masters students. He has received several scientific awards. He has worked as evaluator of projects for international research agencies as well as examiner of Ph.D. thesis for many universities. He is the Editor in Chief of several international journals, Guest Editor of journals, books Editor, book Series Editor and Scientific Advisory for many international journals and conferences. Presently, he is an Editorial Board member of 25 international journals and acts as reviewer for more than 80 prestigious Web of Science journals. In addition, he has also published as editor (and co-editor) more than 100 books and as author (and co-author) more than 10 books, 70 book chapters and 400 articles in journals and conferences (more than 200 articles in journals indexed in Web of Science core collection/h-index 43+/5500+ citations and SCOPUS/h-index 52+/7500+ citations).